Bats

In a World of Echoes

Bats

In a World of Echoes

JOHAN EKLÖF

JENS RYDELL

Library of Congress Control Number: 2018933368

© Springer International Publishing AG, part of Springer Nature 2017
© Photo: Jens Rydell if not stated otherwise
Originally published (in Swedish) by Hirschfeld Förlag 2015
Translation: Neil Betteridge
Graphic design & repro: Göran Andersson
Fact examination: Jeroen van der Kooij
Print & Bound Springer International Publishing AG 2018
ISBN 978-3-319-66537-5

Cover image: Western barbastelle *Barbastella barbastellus*, Italy.

Contents

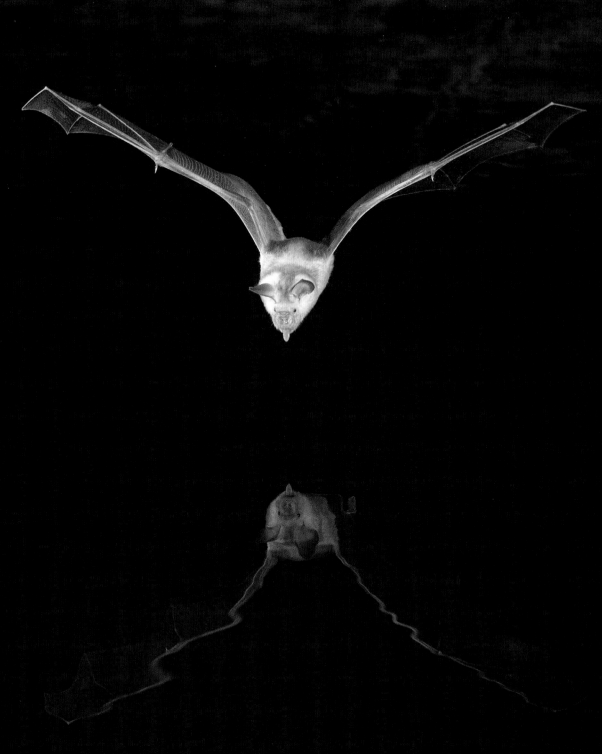

Trident-nosed bat *Asellia tridens*, Israel.

Introduction

"BATS — IN A WORLD OF ECHOES" is about more than one fifth of all mammalian life on the planet. With a timeline dating back to the dinosaurs, it takes you on a journey to exotic locations around the globe and tells a story not just about bats but about a world of sound, about lives lived in darkness, and about mastering the air.

Bats make up their own order of mammals and are in fact more related to carnivores, ungulates, and whales than rodents, contrary to what one might suspect given the old English names "flittermouse" or "rattle-mouse". The modern English "bat" ultimately derives from the Old Norse leðrblakka, which literally translates as "leather flapper".

We currently know about 1300 species of bats and are constantly discovering new ones. They are found over almost the entire planet and are wonderfully diverse in terms of feeding habits and lifestyles, especially in the tropics. There are bats that are fruit and nectar specialists and bats that hunt fish, frogs, birds, and scorpions. In the north, all bats eat insects and other arthropods, while in tropical Latin America there are bats that can live solely on blood, a feat that has earned them—the vampires—considerable notoriety, judging by all the myths and legends that surround them.

Some of the world's smallest mammals are bats, while the largest flying foxes have a wingspan that can reach almost two meters. What they all have in common is the capacity of flight. Bats are the only actively flying mammals, a characteristic that has shaped their appearance more than anything else. Most obvious, of course, are the wings, with the elongated digits that stretch out into flexible, dragonlike wings that propel their lightweight bodies through the air.

Even though bats live in our houses and hunt in our gardens, most people have never encountered a bat up close. We usually see them flitting and darting about, silhouetted against the night sky. Bats evolved under the cover of darkness, and all bats are still nocturnal. There are no markedly diurnal species, as bats spend their daylight hours in the nooks and crannies of trees and buildings or in mines and caves. Their senses—the exceptional night vision of the flying foxes and the echolocation skills of the smaller bats that enable them to create a sonic map of their surroundings—have been honed by the restrictions of nocturnal flight. Being able to navigate and hunt in the dark has made the bats one of the most successful orders of mammals. Yet their aversion to light has often shrouded them in an aura of mystery and fear. In Europe, they are associated with the Devil and Dracula, while in Central America they are the fabled denizens of the underworld. In eastern Asia, however, a different view is held. There, bats symbolise happiness and longevity.

Many bats live dangerously in the human-dominated world, where they must contend with the persecution caused by superstition and the depletion of their forest- and wetland hunting grounds. Other bats, however, are relatively prosperous. Europe's bat populations are currently on the rise, possibly because of a dramatic reduction in the agricultural use of DDT and other pesticides in recent times. While a mere 25 years ago, bats were considered vermin in many European countries, they are now strictly protected throughout the EU, and many people are realising their economic value. Bats perform ecosystem services for multibillion sums every year by pollinating fruit trees, spreading seeds, and eating disease-spreading mosquitoes and agricultural pests. This book aims to bring the bats out of the shadows and demonstrate just what amazing animals they really are.

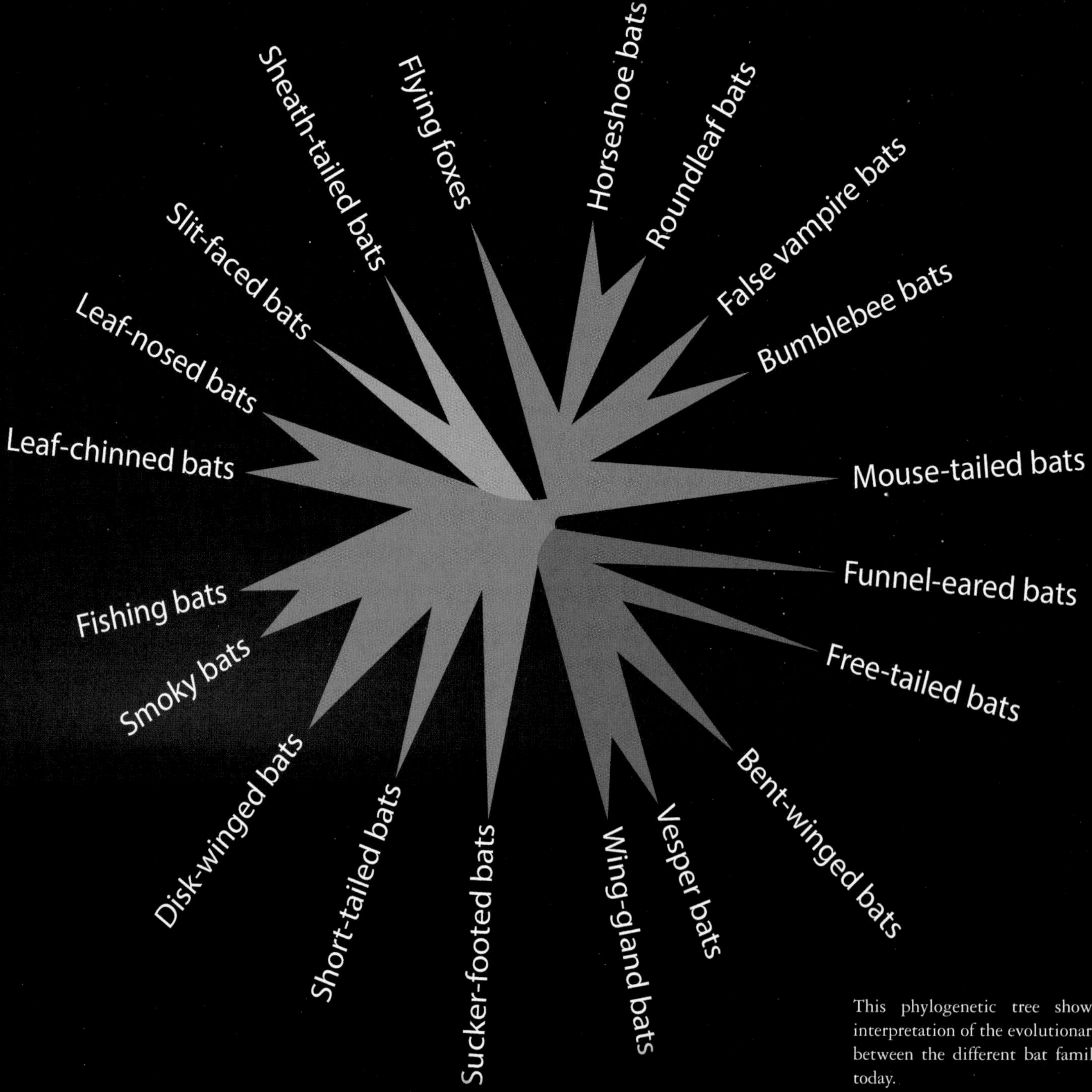

Sheath-tailed bats

Flying foxes

Horseshoe bats

Roundleaf bats

Slit-faced bats

False vampire bats

Leaf-nosed bats

Bumblebee bats

Leaf-chinned bats

Mouse-tailed bats

Funnel-eared bats

Fishing bats

Free-tailed bats

Smoky bats

Bent-winged bats

Disk-winged bats

Vesper bats

Short-tailed bats

Wing-gland bats

Sucker-footed bats

This phylogenetic tree shows a modern interpretation of the evolutionary relationship between the different bat families that exist today.

Evolution and Diversity

Origins

IF WE STEPPED out of a time machine 65 million years ago, at the end of the Cretaceous period, we would find ourselves in a completely unfamiliar world. The seas would be warm and placid, Antarctica would be covered in forest, there would be no Atlantic or Pacific oceans, and Europe would be just one big archipelago. We would be hard pushed to tell herbivores from carnivores and completely unable to point to the ancestors of the shrews or the whales. All these mammals looked pretty much the same, as they were all members of the superorder of Laurasiatheria—one of the four main branches of the mammalian family tree. Amongst these creatures were those that would one day become bats, but since there are no fossil records from them, we have no idea what they might have looked like.

The earliest real bats that we know of—real in the sense that they could fly—lived during the Eocene epoch 50–55 million years ago. Fossil finds from this time, above all those that have been excavated in Messel in Germany and Green River in Wyoming, USA, have provided important pieces of the bats' evolutionary puzzle. The quarry in Messel has become so famous for its fossils that it has been granted world heritage status by UNESCO. The site is teeming with fossils, not only of bats but also of snakes, frogs, fish, horses, and, not least, the primate Ida, who became a global celebrity in 2009 after having hung on the

An early bat as it could have appeared in the Cretaceous period ca 70 million years ago. There are no bat fossils from this time showing what they were like, so we are confined to educated guesses. However, we assume that a wing membrane of some kind already existed, as in the picture, and that it was used for gliding or insect catching.

© Springer International Publishing AG, part of Springer Nature 2017
J. Eklöf, J. Rydell, *Bats*, https://doi.org/10.1007/978-3-319-66538-2_1

wall of a private collector for 25 years. The bats, like most of the fossil remains from Messel, are extremely well-preserved. We are not talking of odd fragment of jaws and skulls but of complete skeletons, some even with preserved soft parts, such as fur and skin, and sometimes even stomach contents.

The Eocene bats looked surprisingly like their modern descendants. The structure of their inner ear gives clear indication that they already had a sense of hearing well-suited for echolocation. Fossilised remains of their stomach contents also show that they ate flying insects, just like the bats of our own time do. The structure of their skeletons also bears witness to a sophisticated capability for flight. All this tells us that bats were airborne echolocators 50 million years ago, exactly like they are today.

Evolution

The bats' elongated digits have earned them their scientific name, Chiroptera, which derives from the Greek words *cheir* and *pteron* meaning hand and wing. Carl Linnaeus observed similarities between the bat skeleton and our own and placed the bat, which he called "Vespertilio", amongst the primates along with humans, apes, and lemurs. Today, however, bats make up their own order that has as its closest relatives in the carnivores, ungulates, and whales.

Bats were previously divided into two main groups, the Microchiroptera or microbats and the Megachiroptera or megabats, these latter often being referred to as flying foxes. Microbats use echolocation to navigate through the dark, while the flying foxes rely more on their sense of smell and their virtually unique nocturnal vision. Given this categorisation, it would be tempting to conclude that the bats' capacity of flight evolved before echolocation, and some scholars have even proposed that the two groups evolved independently of each other. However, according to modern research, genetic similarities show that the flying foxes are merely a family amongst other bats and that all bats have the same ancestor. This means that echolocation either evolved several times in specific branches of the bat family tree

or that it was a primitive feature, later abandoned by the flying foxes. Recently, researchers found similarities in the foetal development in the inner ear between micro- and megabats, supporting the idea that echolocation was an early adaptation in all bats.

At the end of the Cretaceous and the start of the Palaeogene, there was a floral explosion, a rapid evolution of the angiosperms or seed plants that added flowers and fruits to the food chains. This was matched by a parallel and similar explosion of insects. It is likely, then, that fruit, nectar, and insects were important sources of food for primitive mammals, which were often small, crepuscular tree dwellers. The dim light at dawn and dusk puts their powers of vision to the test, and in such a world, it is not hard to understand the benefit of an additional sense. Most mammals have a well-developed sense of smell, good night vision, and sensitive whiskers, but given that many small mammals also communicate using high-frequency sound, the tools were in place for the evolution of echolocation. Even if the step from creating and perceiving sound to being able to interpret echoes is a long one, the simplest possible response is enough to gain the edge on the competition, for the most rudimentary of signal is better than no signal at all. It has been demonstrated that shrews use a simple form of echolocation to avoid obstacles, and by refining this kind of technique, bats could start navigating their airspace and eventually hunt in the dark, guided by echoes, and exploit a source of food inaccessible to other animals, namely, nocturnal insects. Echolocation abilities also spared them having to compete with birds, and, perhaps more importantly, they did not have to emerge during the day and expose themselves to predators that hunt by vision. This is not to say, however, that echolocation evolved before the ability to fly. Indeed, the case could be the converse, and there are several theories about how this could have happened.

Perhaps the bat started off as a gliding animal before taking the step to active flight. The advantage of gliding is that it enables a quick escape from enemies and efficiency of movement between trees. Other

The portraits on the following three pages show representatives from 12 of the 20 extant families of bats.

Noack's roundleaf bat *Hipposideros ruber*, Kenya.

Egyptian tomb bat *Taphozous perforatus*, Kenya.

Hairy-legged vampire *Diphylla ecaudata*, Mexico.

Mexican greater funnel-eared bat *Natalus mexicanus*, Mexico.

Wagner's leaf-chinned bat *Pteronotus personatus*, Mexico.

Mehely's horseshoe bat *Rhinolophus mehelyi*, Portugal.

The African "house bats" occur in different colors and sizes. They are usually hard to identify, however, and although they are common and live near humans, it appears that some of the species have not been described and given scientific names.

"Yellow house bat" *Scotophilus* sp., Kenya.

Dark-winged lesser house bat *Scotoecus hirundo*, Kenya.

White-bellied house bat *Scotophilus leucogaster*, Kenya.

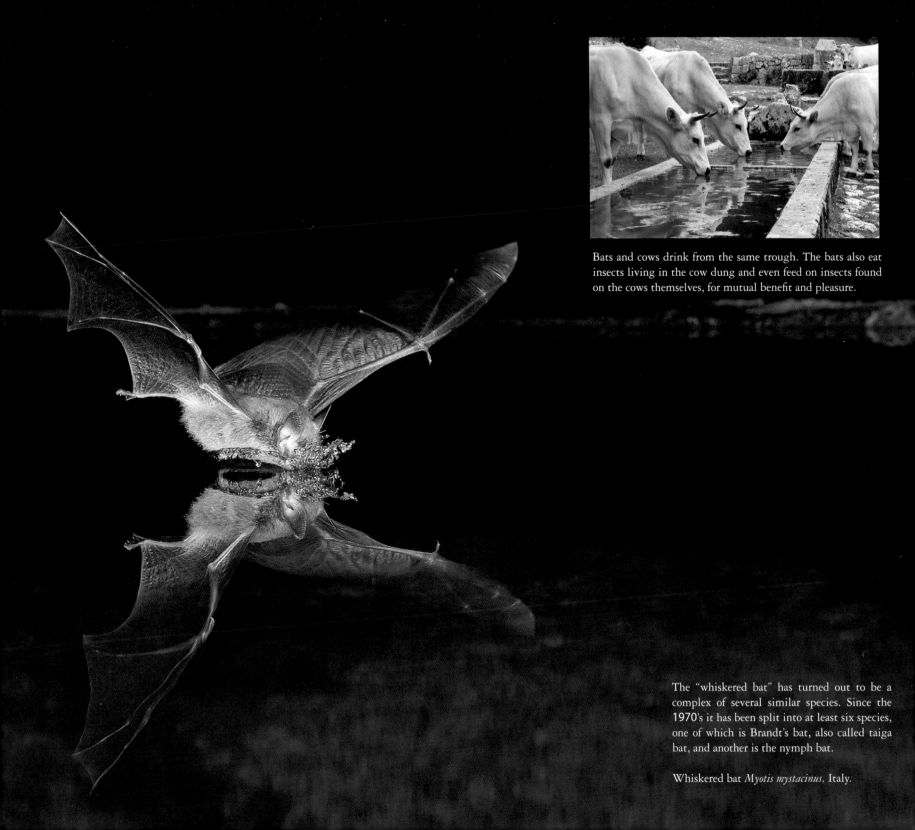

Bats and cows drink from the same trough. The bats also eat insects living in the cow dung and even feed on insects found on the cows themselves, for mutual benefit and pleasure.

The "whiskered bat" has turned out to be a complex of several similar species. Since the 1970's it has been split into at least six species, one of which is Brandt's bat, also called taiga bat, and another is the nymph bat.

Whiskered bat *Myotis mystacinus*, Italy.

Taiwan in the South China Sea is isolated enough so that animals that have reached the island have remained there and evolved in their own direction and in some cases become endemic species. Several of the more than 30 species of bats in Taiwan are endemic, which means that they occur only there. The large roundleaf bat on the picture is one of them.

Formosan roundleaf bat *Hipposideros tera-sensis*, Taiwan.

mammals, like flying squirrels and flying lemurs, have also evolved the ability to glide, as have members of some other animal groups, such as frogs and snakes. However, the step from gliding to active flight is a long and complex one and requires several physiological and anatomical adaptations. Another theory is that the bat wings first evolved as an insect net as their enlarged hands enabled them to scoop up food from the air. This theory is less popular amongst scientists than the gliding theory, but it is interesting to note that many modern bats catch their prey in just this way.

Relationships

We know about 1300 extant species of bats and these are distributed amongst 20 families, although this figure varies depending on definition. One example is the three vampire species that most experts count amongst the leaf-nosed bats but which sometimes are given their own family.

The largest bat family is the vesper or common bats with about 300 species distributed almost over the entire globe. The flying foxes are

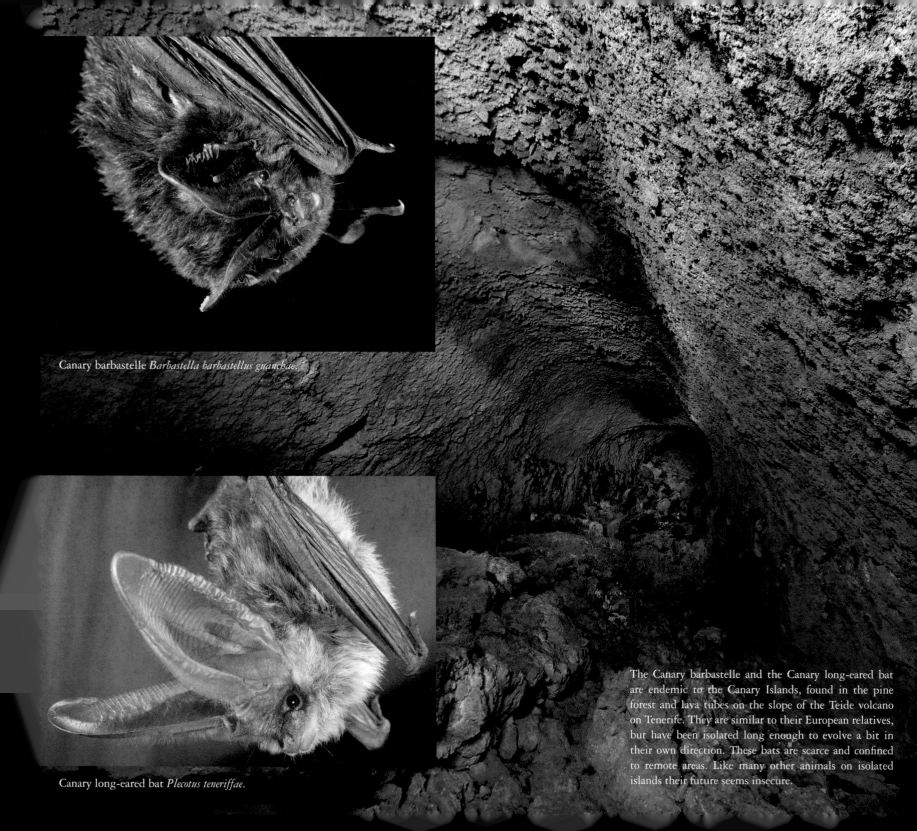

Canary barbastelle *Barbastella barbastellus guanchae.*

Canary long-eared bat *Plecotus teneriffae*.

The Canary barbastelle and the Canary long-eared bat are endemic to the Canary Islands, found in the pine forest and lava tubes on the slope of the Teide volcano on Tenerife. They are similar to their European relatives, but have been isolated long enough to evolve a bit in their own direction. These bats are scarce and confined to remote areas. Like many other animals on isolated islands their future seems insecure.

the second largest family, with around 200 species. The most diverse family is the leaf-nosed bat of the New World, which consists of 150 species from vampires and carnivorous bats to fruit and nectar specialists. Along with the horseshoe bats and free-tailed bats, the above families make up the vast majority of all bats. Other families consist of a few species each, in a couple of cases only one.

Most of today's bat families evolved during the Eocene, but already during the Cretaceous period, 20 million years earlier, there were two established branches of the bat family tree. On one side, we currently find the flying foxes, the horseshoe bats, the roundleaf bats, and a few others, and on the other most of all microbats, including vesper and leaf-nosed bats. It is interesting that the flying foxes, which normally do not echolocate, and horseshoe bats, which are extremely sophisticated echolocators, are more closely related to each other than to other bats. This said the horseshoe bats and their relatives do employ an echolocation technique very different to that found amongst the bats on the other main branch of the family tree.

New Species

The bat family tree is being constantly updated as new, previously unknown species are discovered on expeditions to rainforests or remote caves, although new bat species are also found in Europe now and then as well. Species that we thought unique turn out to be a complex of different distinct populations, or bats we considered rare are discovered in new locations. In a world of scents and sounds, distinguishing characters are not always visible to the naked human eye, and it is only with ultrasonic detectors, DNA techniques, and other modern methods that we can tell certain species apart and understand their lives in the light of kinship. For example, the pipistrelle, which was described in Germany back in the 1700s, recently turned out to consist of two species which are almost identical apart from the frequency of their echolocation signals and some genetic differences. In Europe alone, a dozen new bat species have been described in the past few decades with the aid of DNA sequencing and careful examination of recorded ultrasound signals.

Distribution

In northern Norway, way above the Arctic Circle, live the world's northernmost bat colonies. The summer months there offer an abundance of food, but the season is short and the window of opportunity these bats have to reproduce is very narrow. The closer we come to the equator, however, the more species we find established. In southern Sweden, there are 19 species of bats, in Europe there are roughly 50 species, in a tropical country often about 100, and in the entire tropical America around 200. Only three families, the horseshoe bats, vesper bats, and free-tailed bats, are common in temperate areas. Polar regions and oceans excepted, bats can be found in all conceivable environments, and many species range over vast areas, and while some fly over several tens of kilometres on the hunt for food, migratory species can cross several 1000 km between their summer and winter homes. On some of the planet's most remote places, like Hawaii and New Zealand, bats are the only native mammals. Evolution in isolation has resulted in new species adapted to the particular conditions on the islands.

Despite their ability to fly and thus cover long distances, many bats are surprisingly homely and rarely, if ever, venture any further than a few kilometres from their roosts. Such is the way, for example, of the brown long-eared bat in Europe, which often keeps to the same garden or church all year round and sometimes for life. This species, like many of its stationary kin, especially in the tropics, are ill-disposed to flying over open terrain and therefore become isolated when the surrounding habitats are destroyed.

Greater horseshoe bat *Rhinolophus ferrumequinum*. England.

Roger Ransome and the horseshoe bats

In the old quarry at Comb Downs, which once provided building material for the city of Bath, two species of horseshoe bats spend the winter. Roger Ransome has followed these bats carefully for more than 50 years. All individuals of the greater horseshoe bat were given numbered metal rings on the forearm when they were young and they have subsequently been checked annually for growth, health and reproductive success. They have become used to this activity and consider Roger's visits as a normal part of life. These bats have no secrets. Their lives are known in detail, including their health problems and social relationships. Together with Roger the bats provide important information about the development of a wild mammal population over a long time, ever since the agronomical revolution and the introduction of pesticides in the mid 1900's to the start of the great climate change.

The air is suddenly full of wings, which like a bee-swarm appear from the dark in the ceiling of the old weapon store, a rusty remain from the Second World War. Within a moment after dusk the chase for tiny insects has begun among the fresh green vegetation, following the first rains. The smallest insect-eating bats are super light-weight aerial acrobats.

Northern cave bat *Vespadelus caurinus*, Australia.

Flight and Morphology

THE ABILITY TO FLY and locate food in the dark is what has made the bat's morphology and function so special, from its size and metabolism to the shape of its sensory organs. The leaf-nosed bat's nasal outgrowths, the horseshoe bat's peculiar physiognomy, and the large eyes of the flying foxes are all examples of how bats have become adapted to a life in dim light.

Bats are generally small and, above all, light. Most are about 4–10 cm long with a wingspan five times their length and weigh less than 20 g. While lightness is primarily an advantage for flying, echolocation also requires its physical and physiological adaptations. High-frequency sound rapidly attenuates in the air and reveals obstacles and prey for only a fraction of a second, and the information disappears just as quickly as it appears. Insect-eating bats in particular also need to be extremely manoeuvrable, able instantly to react with acrobatic deftness as soon as an echo is detected. The smallest bats are also some of the smallest mammals, weighing in at only 2 or 3 g, less than one page of this book.

Flying foxes, which are fruit- and nectar-eating bats found only in the Old World, are on average much larger than the leaf-nosed bats, which are their New World fruit- and nectar-eating cousins. This is an evolutionary adaptation it would seem, to their native fruits, which are larger than those of the New World. But there are other and perhaps more likely reasons for the size difference between the fruit-eating bats of the Old and New Worlds. Some of the bat predators of the New and Old Worlds are different. In the New World, bats resting on a flower or a fruit can be captured by arboreal pit vipers, snakes that can detect the heat emitted from the bat even in total darkness, and strike immediately.

Because of these sit-and-wait predators, bats take precautions either by picking the fruit and carrying it away to another spot or hover in front of the flower. Hovering requires extreme manoeuvrability which in turn calls for a small and light body. In the Old World, however, there are no pit vipers, so there is no reason for bats to hover in front of flowers or fruits in the dark. Instead they sit down and eat in peace and therefore do not need to be particularly light and manoeuvrable.

Wing

The most striking characteristic of the bat compared with other mammals is, of course, the wings. Greatly elongated hand digits lend stability to an elastic membrane, the patagium or wing membrane. The index finger constitutes the leading edge, the middle finger its end point and hence the wingspan, and the ring and little fingers its width. Together, the fingers and skin form a neat umbrella-like structure. The only digit not part of the wing proper is the clawed thumb, which projects from the wing and which the bat can fold down when flying slowly to generate greater lift. Otherwise, the thumb is used for climbing and scratching, as well as for locomotion on the ground. Flying foxes, unlike other bats, have a claw on the index finger as well. Vampire bats have a particularly well-developed thumb, which they use in combination with their feet and wings to launch themselves off the ground and into the air.

Flying puts extraordinary strain on joints and bones. To prevent too much load on the skeleton, all finger joints are angled differently relative to each other, and closest to the fingertips, the bones contain less calcium, which makes them extra flexible. Bats not only have wings on their

© Springer International Publishing AG, part of Springer Nature 2017
J. Eklöf, J. Rydell, *Bats*, https://doi.org/10.1007/978-3-319-66538-2_2

The bat's hand has the same five fingers as our hand. The wing consists of elastic skin forming a membrane with blood vessels, muscle fibres in bundles, tendons and thin flexible finger bones.

Daubenton's bat *Myotis daubentonii*, Sweden.

Egyptian tomb bat *Taphozous perforatus*, Kenya.

The large flying foxes usually live in colonies that move
around over extensive areas. They sleep in the open under
the burning tropical sun, and use the large, thin wings as

"hands", their entire body serves as a wing. Their hip bones are turned away from the body, and the wing membrane stretches between hands and feet like a paper kite. The wing membrane makes up 90% of the bat's body surface area and comprises two epithelial layers with nerves, blood vessels, elastic fibres, and muscles in between. The surface is coated with sensitive hairs connected to nerves that transmit information to the brain about the air flow over the wings. This enables bats to constantly make fine adjustments to their flight path and manoeuvre optimally through the air. The bat wing membrane is one of the most rapidly healing structures in mammals. Rips and small holes usually heal within a week, and while larger wounds can, of course, be devastating to the ability to fly, long tears and even bone fractures usually heal before the bat starves to death.

The relatively large surface area and rich blood supply of the wing membrane allow the wings to regulate the bat's body temperature. Warm blood from the body is cooled when pumped through the thin membrane, reducing the risk of overheating from the intense energy metabolism that flying entails. In tropical areas, bats can sometimes be seen flapping their wings while hanging in trees resting. This they do to rid themselves of surplus body heat, in much the same way as elephants flap their ears.

Tail

Most bats also have a similar membrane, called the uropatagium or tail membrane, stretching between the feet and enclosing the tail, which gives extra lift at low speeds and both controls the angle of attack—the angle between a wing's leading edge and the oncoming airflow—and serves as an air brake. The tail membrane also forms a net or bag for trapping insects, and around midsummer when the nights are at their shortest, a female bat, hanging upright from her thumbs, might wrap it around her sleeping pup like a papoose.

Some bats have a tail and tail membrane almost as long as the rest of the body, while others, such as the free-tailed and sheath-tailed bats, do not use the tail membrane as an insect catching bag to the same extent but have a free, protruding tail instead. One bat family, the mouse-tailed bats, have found a unique area of application for their extremely long tail. They use it to probe their way along as they crawl backwards in caves and amongst rocks. Vampires, which are partly terrestrial, have almost no tail and a very short tail membrane, which frees up the legs for jumping and running.

Face

If the morphological properties of the bat body are determined by the challenges of flight, the faces, especially those of the insectivores, are shaped by those of echolocation. Many bat species have strange, striking physiognomic features, like leaf-like nasal protuberances, enormous ears, projections, and furrows. We still know very little about how these adaptations benefit echolocation, but presumably all of them are intended, in different ways, to direct sound, amplify particular frequencies, filter out unnecessary information, or facilitate the detection of weak signals. In species that listen for the sounds of prey animals, the length of the ears can sometimes equal that of the body, as is the case with the long-eared bats, which can pick up the sound of a moth's steps on a leaf. One distinctive ear feature of echolocating bats is the fleshy projection by the auditory canal called the tragus. While in humans, this structure is fairly discrete, in bats it can be anything from a long, narrow, fairly flexible flap to a broad, rounded lobe. The tragus helps the bat to localise the source of echoes vertically, while horizontally, they are localised through the stereo effect created by having two ears. A recently discovered property of bats is the mutability of the outer ear—the pinna—the shape of which they change in a fraction of a second to catch as much sound as possible and optimise the perception of sonic nuances.

Most bats use their mouth for emission of the echolocation calls, adjusting the lips to direct and concentrate the energy of the sound. Many bats, such as the long-eared bats, use their nose for echolocation, while the barbastelle uses both. But above all, it is the leaf-nosed bats, the roundleaf bats, and the horseshoe bats that use their noses to emit

Northern bat *Eptesicus nilssonii*, Sweden.

Greater mouse-tailed bat *Rhinopoma microphyllum*, Israel.

Silky short-tailed bat *Carollia sarelli*, Belize.

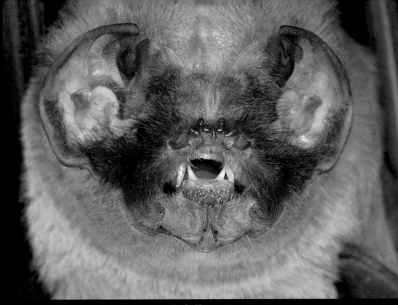

Ghost-faced bat *Mormoops megalophylla*, Belize.

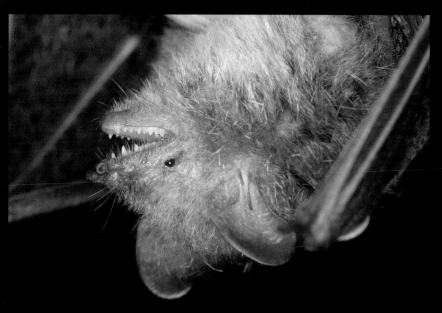

Faint-golden little tube-nosed bat *Murina recondita*, Taiwan.

Great fishing bat *Noctilio leporinus*, Belize.

Greater mouse-eared bat *Myotis myotis*, Portugal.

Peter's dwarf epauletted fruit-bat *Micropteropus pusillus*, Kenya.

Heart-nosed bat *Cardioderma cor*, Kenya.

Andersen's slit-faced bat *Nycteris aurita*, Kenya.

Black bonneted bat *Eumops auripendulus*, Belize.

African trident-nosed bat *Triaenops afer*, Kenya.

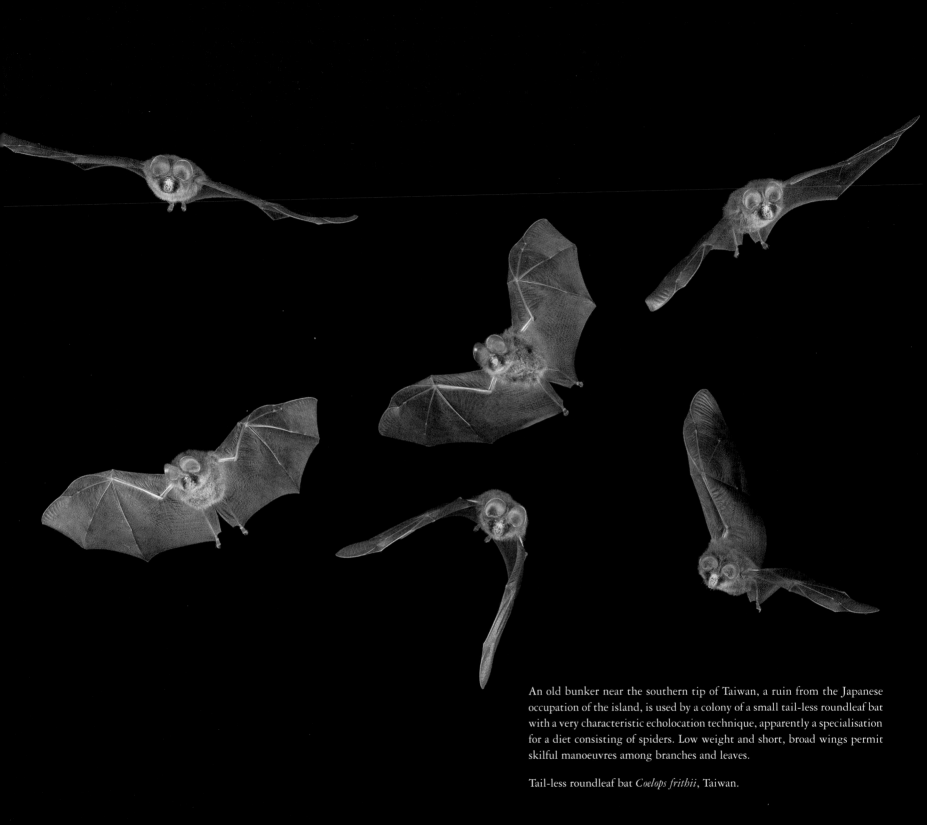

An old bunker near the southern tip of Taiwan, a ruin from the Japanese occupation of the island, is used by a colony of a small tail-less roundleaf bat with a very characteristic echolocation technique, apparently a specialisation for a diet consisting of spiders. Low weight and short, broad wings permit skilful manoeuvres among branches and leaves.

Tail-less roundleaf bat *Coelops frithii*, Taiwan.

White wings seem to be a poor adaptation in bats, but it is nevertheless found in some tropical species. Perhaps it is functional at dusk and dawn, when it may be a camouflage against the evening sky?

Hildegarde's tomb bat *Taphozous hildegardeae*, Kenya.

Some cave-dwelling bats in the tropics are bright orange, but the colour varies for some reason a great deal from cave to cave even when the species is the same.

Noack's roundleaf bat *Hipposideros ruber*, Kenya.

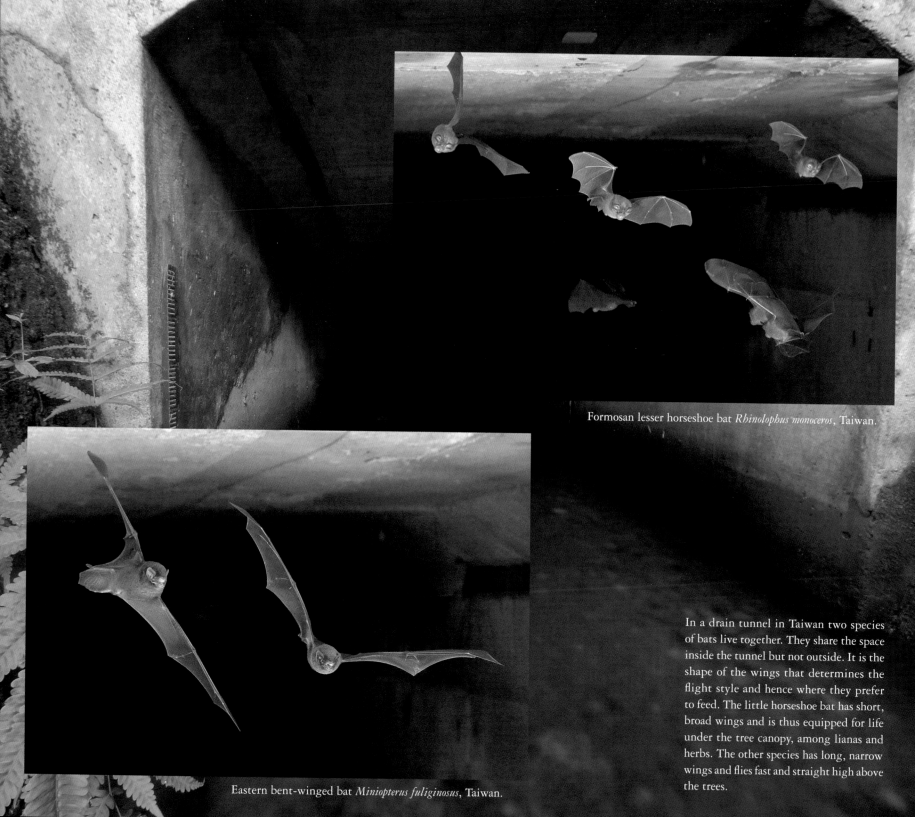

Formosan lesser horseshoe bat *Rhinolophus monoceros*, Taiwan.

Eastern bent-winged bat *Miniopterus fuliginosus*, Taiwan.

In a drain tunnel in Taiwan two species of bats live together. They share the space inside the tunnel but not outside. It is the shape of the wings that determines the flight style and hence where they prefer to feed. The little horseshoe bat has short, broad wings and is thus equipped for life under the tree canopy, among lianas and herbs. The other species has long, narrow wings and flies fast and straight high above the trees.

Daubenton's bat always shows up at the old stone bridge after dusk. Its specialised hunting technique over calm water surfaces requires that the focus is straight ahead and that the wing beats are shallow enough to get free of the water surface.

Daubenton's bat *Myotis daubentonii*, Sweden.

ultrasound for echolocation, and it is mainly in these bats that we find facial growths or projections, structures that in one way or another direct or amplify their sonic pulses. The horseshoe bats have narrow furrows along the nose that serve as resonance chambers to amplify certain frequencies, not unlike the bodies of wind instruments.

Colour

Bats live in a world of sound in which auditory impressions usually are more important than visual ones and in which the shape and smell of the body are more important than its colour. Most bats have a coat of various shades of greyish brown, although some species do stick out with their buff- or orange-coloured fur or dappled, striped, or spotted patterning. Some small flying foxes have distinct yellow- and lime-coloured spots against brown wings. The patterns are individual and probably serve as camouflage amongst dry leaves of diverse colours. Most of the bats with some kind of prominent colour or pattern are species that roost in trees and bushes. In Central America lives a snow-white leaf-nosed bat that seeks refuge under large leaves, where the tropical daylight tints its fur green and thus prevents any telltale shadows from showing through. The light colour therefore serves to camouflage it amongst the vegetation. Other bats are more even-coloured, and many are lighter on their bellies than on their backs, just like seabirds and Second World War aircraft. The purpose is, of course, the same, to be as invisible as possible when flying from above against the dark ground and from below against the light sky. Bats that have really mastered the art of invisibility are some African and South Asian species, the white or sometimes transparent wings of which filter the twilight and blend with the colours of the evening sky.

Honduran white bat *Ectophylla alba*, Costa Rica.

The white bats live in small permanent groups. During the day they cluster under *Heliconia*-leaves, immobile and difficult to detect, in the tropical rain forests of Central America. The male in the group makes the tent by biting along the central nerve of the leaf, which makes it collapse and form a protective roof.

Brown long-eared bat *Plecotus auritus,* Estonia.

With the ears bended under the folded wings, a brown long-ear hibernates in a cellar, hanging in a root in its curved claws.

Flight Style

When flying, bats stretch out their wings with their palms facing down and their thumbs pointing in the direction of their flight path. They press their wings down in a distinct manoeuvre and with a flick of the wrist twist them upwards again. Powerful shoulder muscles bring the wings over the head and in towards the body before stretching them out once again. The sequence is then repeated, with the bat engaging its entire body to create a flap rate of about ten per second, depending on its size.

Bats and birds are differently equipped for flight. Instead of a coating of separate feathers, bat wings have an elastic membrane, and instead of only two joints, the bat has as many as in the human hand, allowing them to twist and turn their wings more than birds can. Bat flight is therefore normally more energy demanding than that of birds and less conducive to gliding. Bats also fly more slowly, rarely more than 30–40 km/h, although they are able to fly faster, at an increasing energy cost, with some species able to reach 100 km/h or even more in an emergency. Nevertheless, it is usually less important for a bat to fly at speed than it is to manoeuvre and react with lightning agility, which it does partly by moving each wing separately in unsynchronised movements.

The shape of the bat wing is an indication of how swift and manoeuvrable they are, and from this we learn a great deal about their lifestyles and ecological niches. Bats with short, broad wings are relatively slow but masters at navigating the narrow-cramped spaces between branches and leaves or in barns and bell towers. Short, pointed wings often make for a faster yet slightly stiffer flying technique and are better suited to larger open spaces, such as in glades or along forest paths. Species with long, narrow wings hunt in wide-open spaces, above the tree canopy, over lakes, or in open terrain, where they are less likely to encounter obstacles. The long wings are more energy efficient but provide less manoeuvrability. These bats are often, but not always, fast flyers.

Many species, such as the northern bat, hunt a few metres above ground level and patrol back and forth over meadows and fields, aiming their echolocation pulses diagonally downwards so that any insects thus discovered can be caught with a sudden perpendicular strike. By hunting downwards like this, the bat can make use of gravity to gain extra momentum in its attack. When its prey is discovered, the bat rolls upside down and, beating its wings upwards, spins round and swoops, not unlike a dive bomber. Birds, like swallows, that catch insects in the air are unable to see their target against the ground and have to search against a lighter background, normally the sky. This means that instead of diving, they must catch their prey from below or possibly from the side, although they do have the benefit of daylight to spot it in the first place. Bats that hunt just above water surfaces, which are called trawling bats, such as Daubenton's bat and fishing bats, cannot, of course, exploit the diving technique and have to attack from the side or upwards to avoid getting wet. The echolocation pulses of such bats are directed straight ahead, parallel with the surface of the water.

When catching its prey, the bat stops, apparently mid-flap, curls up, and tucks its legs and tail into its body to enclose its meal. Bats often use their tail membrane, to catch their prey like this, and sometimes their wings as well. An instant after the moment of capture, the wings are back in motion and the ultrasonic calls are once again targeting the next mayfly, beetle, or mosquito. Every dart, every duck, and every dive are achieved with the bat's flexible wings and tail membrane acting in harmony to control its speed and fine-tune every manoeuvre.

On the Ground and at Rest

The bat's mode of rest has had to adjust to its life in the air. Since their legs are more or less part of their wings, it is easiest for bats to hang upside down from their feet or cling onto a wall with both thumbs and feet. Bats either land on all fours and then crawl into a resting position, or they do a somersault and let their backward-pointing feet grab onto the underside of a roof and lock against it. Although their heads are pointing down, an unusually mobile neck means that they can keep their eyes pointing ahead. Their legs are also twisted so that they can hang with their bellies flat against the wall while keeping a watchful eye on the space around them.

Most bats hang from walls or hide in crevices. Some, however, such as flying foxes and horseshoe bats, always hang freely by their feet from branches or ceilings with their wings wrapped around their bodies so that they can gain lift as soon as they let go. Hanging free is an effective means of avoiding predators, and, because the bat's droppings accumulate on the ground far below, it is also hygienic. Brown long-eared bats also hang freely from ceilings, especially during their winter hibernation.

Most bats are reluctant to land on the ground since it increases the risk of capture by a predator. While a healthy individual rarely has difficulty launching itself into the air from ground level using its wings and tail, a sick or tired bat can have problems. As can wet bats after accidentally ending up in water while drinking or catching insects, and

Sundevall's roundleaf bat *Hipposideros caffer*, Kenya.

To hang with the claws against the rock and with the head and attention in the opposite direction requires flexible legs and neck.

even though they can swim well, they need time to dry off, as a drenched coat impedes agility and wastes energy.

Some species, however, have developed the art of walking so well that they actually spend much of their time on the ground. Vampire bats, for instance, usually land some distance from their victims and cover the last few metres scuttling and skipping towards them. When taking off again, they use their powerful thumbs, wings, and feet to propel themselves upwards. New Zealand's lesser short-tailed bat has taken yet another step back to being a quadruped. They have a long history of evolution on the geographically isolated islands and, for much of the time, were their only mammal species. They were therefore free to develop a unique—for bats, at least—hunting technique, seeking out food amongst the fallen leaves and flowers. Long-eared bats and certain mouse-eared bats are also known to hunt insects on the ground, such as flies and beetles crawling on fresh cow dung.

Senses

PLATO BELIEVED THAT humans experience an impoverished reality filtered through the senses. If so, we will therefore never be able to understand what it is like to be a bat. However, even though it is impossible for us to fully relate to the experience of ultrasound or infrared, we are able to build and use instruments with which we can detect the many sounds that we cannot hear otherwise, such as the squealing of tram brakes, the chirping of grasshoppers, or the bending of grass underfoot. For bats, the ability to extract valuable information from high-frequency sounds and use echolocation to build up sonic images of their surroundings has made them masters of the dark. It is, if you will, the chiropteran sixth sense.

"Seeing" with the Ears

Back in 1793, Italian scientist Lazzaro Spallanzani and his Swiss colleague Charles Jurine demonstrated that while blindfolded bats were able to negotiate an obstacle course, bats fitted with earplugs could not. They concluded that bats could "see" with their ears. However, the prevailing view amongst zoologists at the time was that bats had particularly sensitive skin on their wings that allowed them to navigate in the dark by feeling their way through the air. Spallanzani, who was known for his clever biological experiments, met with little sympathy and his ideas were not accepted, not until 1938.

To fully enjoy the bats some modern high-tech equipment is needed, such as ultrasound detectors and IR-video. Toni Guillén-Servant dressed up for field-work in Belize.

The then 23-year-old Harvard student Donald Griffin had read that bats use sounds at a frequency higher than the human ear could perceive. These ideas were posited after the First World War when the defence industry was experimenting with sonar, a sound-based technique for tracking submarines. Griffin contacted a physicist who had built a device that could make high-frequency sounds audible to humans, the first ultrasound detector. He visited the physicist with some bats, which he released into the room, and shortly afterwards, the pair of researchers coined a new term—echolocation.

At roughly the same time, similar discoveries were being made in Holland by Sven Dijkgraaf. However, the Second World War had just broken out in Europe, and he was unable to publish his work until the war was over. Griffin and Dijkgraaf had both independently discovered that bats emit high-frequency calls and listen for the echoes in a way analogous to how engineers conceived the mechanisms of submarine sonar. The principle is the same as when we shout across a valley and are met by an echo bouncing back from the other side. But where we hear a mountain, bats can hear a caddisfly. Spallanzani and Jurine were thus quite correct, bats can indeed "see" with their ears. But before we explore how they do this in more detail, we must define what sound is.

Perceiving Sound

If you pluck a guitar string, its vibrations excite the air around it, causing it to contract and expand and form sound waves that travel at a speed of around 340 m/s, depending slightly on the ambient

© Springer International Publishing AG, part of Springer Nature 2017
J. Eklöf, J. Rydell, *Bats*, https://doi.org/10.1007/978-3-319-66538-2_3

temperature and humidity. When these sound waves strike the tympanic membrane or ear drum, it starts vibrating at the same rate. These vibrations are then converted in the inner ear into nerve impulses, which are transmitted along the auditory nerve to the brain, where they are interpreted as sound. The frequency of the sound, measured in hertz (Hz), is the number of waves passing a single point in 1 second, the shorter the waves, the more of them will pass. A low-frequency sound thus comprises few but long waves and is heard as a low-pitch tone, while a high-frequency sound has many yet short waves and is heard as a high-pitch tone.

There is a formula for this, which can be used to calculate the frequency of a certain wavelength or vice versa; speed (340 m/s) = wavelength (m) × frequency (1/s).

The most acute human ear can hear a frequency spectrum ranging from approximately 20 Hz to 20 kHz or from 20 to 20,000 waves a second. Frequencies higher than 20 kHz are referred to as ultrasound. The tiny and, above all, thin tympanic membrane of the bat can vibrate much faster than our own, enabling it to pick up higher frequencies of up to 150–200 kHz.

Low-frequency sound has a longer range than high-frequency sound, a phenomenon that we experience, for example, when our neighbours are throwing a party and all that reaches us through the walls is the seemingly relentless baseline of the music. The treble—the light, shortwave sounds—quickly dissipates. The volume or amplitude of the sound is measured in decibels (dB) and depends on how hard we pluck our guitar string. The more force we use, the more energy we transfer and the higher the amplitude. High amplitude or loud sounds travel further. The lowest amplitude or quietest sound we can hear, theoretically, is 0 dB, while sounds over 130 dB can cause physical harm to our hearing.

What we call a tone actually consists of several subtones. At the lowest frequency is the fundamental, over which are layered harmonics at, normally, double, triple, and quadruple this frequency. A sound of 10 kHz has harmonics of 20 kHz, 30 kHz, etc. This applies as much to musical instruments as to human and bat voices. The strengths or relative amplitudes of the different subtones create the instrument's or voice's timbre, or its typical character, and depends on how the air vibrates or resonates within the instrument or within the mouth, nose, and sinuses.

Echolocation

Echolocation is the use of sound or radio waves and their echoes to detect and locate objects. In the former case, the technique is also called sonar, which is an acronym for *so*und *n*avigation *a*nd *r*anging, and in the latter case radar, from *ra*dio *d*etection *a*nd *r*anging. Sonar is a biological as well as a technical term, since it is used by both animals and men, while radar is purely technical.

Sound reflected off an object can contain information about its distance and size, as well as its movements and speed. However, for a clear echo to be obtained, the wavelength of the sound must not be longer than the size of the object being searched for. A mosquito has a wingspan of about 1 cm. So, for a sound to rebound off a flying mosquito, it needs to have a wavelength of 1 cm or less. A short wavelength is the same as a high frequency, and 1 cm is equivalent to 34 kHz, according to the formula above, which is well within the ultrasonic range. It is thus the diet of small insects that led to the evolution of the bat's use of high-frequency sound and vice versa.

High-frequency sound, however, has the important disadvantage of being attenuated rapidly in air and therefore only works effectively at distances of a few metres or less. Lower frequencies travel further but only reflects against large objects. In practice, then, the frequency used by the bat when echolocating is a compromise between its need to find small insects and to detect them early enough for a successful catch. The art of navigating and hunting for food with a combination of sound and aerial acrobatics is the secret of the bat's success among mammals. But they are not alone in their ability to echolocate. A sophisticated variant of the technique is also employed by dolphins, while more rudimentary versions are found in shrews, some tropical cave-dwelling swifts and the South American oilbird, which, like bats, live in caves.

Some blind people also use a simple form of echolocation, having developed the ability to use the echoes produced by rapid clicking sounds to negotiate their surroundings with the aid of a stick. Some have even learnt to make their own very bat-like clicks with their tongue. These days more advanced technical aids are also available for this purpose, very much thanks to our increasingly detailed understanding of how bats use ultrasound. Bat echolocation is based on transmitting a series of short, powerful high-frequency pulses and listening for the echo in the short silent periods between. Bats estimate the distance to objects in their path by measuring the time delay between the emitted pulse and the echo. The sooner the echoes return, the closer the object.

Hunting Insects

If it had merely been a question of orientation and navigation, low-frequency sound would have worked just as well as a high-frequency one or perhaps even better. However, it is the hunt for insects that poses the greatest challenge, and that has led to the evolution of the bat's sophisticated echolocation technique using ultrasound. Bats that hunt insects on the wing almost always have to deal with the problem that their prey can only be detected at very short distances, which in turn requires extreme reaction times and aerial agility. In effect, an insect the size of a mosquito may be detected a metre away, which means that if a bat is approaching at a speed of 10 m/s, it has only one tenth of a second to react, home in, and attack, assuming that the insect remains stationary in the meantime, which, of course, it does not. So, the bat must be able to judge the insect's flight path and estimate where it will be at the moment of capture. This is a complicated calculation, but means that the bat usually has a little more time to act than this simple example might suggest.

At any rate, the bat must act on reflex and assume that everything moving around in the air is edible. It has no time to think and take decisions and will therefore even respond to a small tossed stone. It is the need for lightning strikes that explains why insectivore bats are so small and relatively slow. The slower the bats fly, the more time they have to react, switch course, and swoop—although this does make it difficult to intercept insects flying at speed.

All insect-eating bats create sound using their vocal cords, just like other mammals, including ourselves, and transmit it through the mouth or nose, or sometimes both, depending on species. So, the fact that most bats fly with their mouths open and teeth exposed has nothing to do with their wanting to appear ferocious and everything to do with echolocation. Their vocal cords make it possible to form tones, which means that the sound energy can be concentrated to just a handful of frequencies that correspond to their optimal auditory range and that the frequency can be varied as needed. The throat, nose, and mouth serve as resonance chambers, strengthening the tones produced by the vocal cords and either amplifying or dampening the various harmonics. The frequencies used are usually similar amongst individuals within a species, but differ to a greater or lesser degree between species. This means that it is possible, to some extent at least, to determine a bat's species by the frequency of the sounds it produces. Generally speaking, the larger species use lower frequencies to hunt over greater distances for larger prey. Higher frequencies enable bats to also detect smaller animals, which are usually much more abundant, but they also shorten their range and make greater demands on manoeuvrability. The highest frequencies are therefore best suited to the smallest species. There are, however, many exceptions to this general rule, such as the horseshoe bats, amongst which even the larger species are high-frequency echolocators.

Generating a clear echo requires not only a high-frequency sound but also sufficient amplitude for the energy to make it all the way to the target and back. Bats use their lips or nose with its various fleshy projections as megaphones to direct and concentrate the sound energy, which can thus commonly reach levels as unbelievably high as 120 dB, about the same as a thunder clap. Some bats can even produce sounds exceeding 130 dB, measured 10 cm from the mouth. It is therefore fortunate for us that we are unable to hear their sounds. Bats amplify their sounds by engaging their flight muscles to increase the pressure behind their vocal cords, so the interval between pulses, or the rhythm of their

These two bats are Mexican examples of species that hunt insects in the open air. The upper is large, flies fast and uses lower frequencies, searching for larger insects over longer distances. The lower one is small and more manoeuvrable and searches for small insects over shorter distances, using higher frequency calls.

Black mastiff bat *Molossus ater*, Belize.

Black-winged little yellow bat *Rhogeessa tumida*, Mexico.

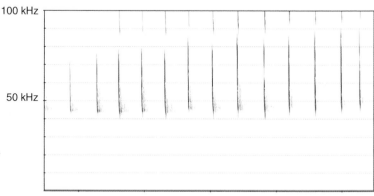

A diagram that illustrates how the frequency of a sound changes with time is called sonogram or spectrogram. Those in this section show what happens in one second but the frequency scale varies between the figures.

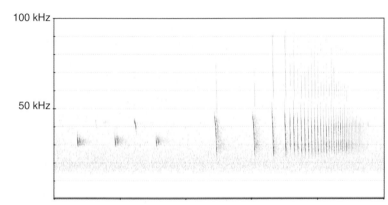

The barbastelle searches for prey using two different types of pulses alternating at different frequencies. Like most bats search pulses are replaced by attack pulses as soon as an insect is detected. This happens about halfway through the sequence. The attack pulses are used for ranging (determination of distance) and therefore consist of steep sweeps that successively gets shorter and more frequent as the insect gets closer.

Western barbastelle *Barbastella barbastellus*, Sweden.

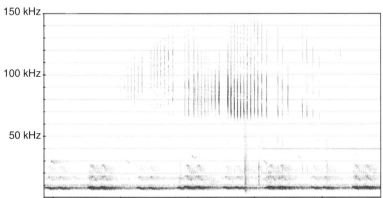

The slit-faced bats feed on aerial insects but also pick those that do not fly, listening to the prey or using echolocation. However, here it uses steep sweeps to negotiate the emergence hole at the roost. Such pulses are somewhat similar to attack pulses.

Cape long-eared bat *Nycteris thebaica*, Kenya.

sounds, usually reflects their wing beats. A pulse every one, two, or three flaps is normal, depending on how far ahead they need to aim, which depends in turn on what they are hunting. However, when a bat determines the exact distance to an object, such as an insect immediately before the moment of capture, the calls must be repeated much faster than the wing beat rate in order to update the exact position of the prey.

Short pulses interspersed with periods of listening for echoes create an aural world analogous to our visual world seen through strobe lighting. The echoes are interpreted individually but are combined into a moving sequence or image in the brain. The pulses must always be kept separate from the echoes, as not even a bat can perceive weak echoes while producing sounds at the same time. This is a fundamental problem for all echo-locating animals and requires each pulse to be short enough not to clash with the echo on its return. Not only this, but the weak echo returned by an insect has to reach the ear before the much stronger echoes from the background. So, when a bat needs to search at close range, it must shorten its echolocation calls to avoid overlap with the echoes, and when searching over a long distance, it must increase the interval between pulses to give time for the echo to return before emission of the next pulse.

Overlapping sounds can also easily create false echoes, which can deceive the bat into believing an object is there when it is not. To mini-mise the danger of misinterpreting overlapping echoes, at least some species build up mental images of every single call, and to distinguish between them, the frequency of each one is fine-tuned up or down. When a noctule hunts, it often gives off a plip-plop-plip sound as the pulses alternate in frequency so that the bat can distinguish between the echoes from each pulse.

Frequency Modulation

It takes precision aiming to locate flying prey, the position of which has to be ascertained with centimetre accuracy, a skill that requires extremely short echolocation calls to avoid a blurred picture. But given that it takes time to maximise the pressure behind its vocal cords, the amplitude of its pulses necessarily decreases as the calls get shorter. Consequently, long pulses tend to be too imprecise and short pulses are too limited in range.

The solution to this dilemma is called frequency modulation. A fre-quency-modulated pulse slides rapidly over many frequencies, from light to dark tones or vice versa. Every little part of the pulse thus has its own frequency, and such pulses are called frequency sweeps or just sweeps. Sweeps can be steep or shallow depending on how fast and extensively the frequency is changed. By measuring the time between the transmit-ted pulse and incoming echo at a certain frequency, the bat is able to accurately judge distance without having to resort to overly short and thus weak pulses. Every individual frequency is heard for a very brief time relative to the pulse as a whole, and the steeper the sweep, the greater the precision. Compared to our ears, those of the bat are extremely well-adapted to distinguishing between frequencies and separating them out before transmitting the auditory signals to the brain. Frequency modulation is thus an essential tool for bats when it comes to judging the exact position of small, highly mobile prey animals.

When hunting on the wing in the open air, a bat normally needs longer pulses despite their unsuitability for judging distance. A flying insect flaps its wings at a certain rate, and smaller insects flap their wings faster than larger ones. At the same time, an echolocation pulse only generates a clear echo when striking a flapping insect at the precise moment when the wings are outstretched and roughly perpendicular to the direction of the sound waves. All other positions return a less distinct echo. Hence, the flapping, undulating echo of a flying insect helps the bat to locate it and considerably improves its sonar range. It also allows the bat to estimate the size of the insect and sometimes even the species. But this requires the bat's pulses to be at least as long as the insect's wing beats—effectively 5–20 ms—and of approximately the same frequency throughout.

Search pulses, which the bat uses to find insects, are thus often rela-tively long and contain few frequencies, shallow sweeps, while attack pulses, on the other hand, which are essential for judging distance, are shorter, more frequency-rich, steep sweeps, and also quicker in rhythm.

Early summer at lake Vättern in Sweden. There is no wind and a completely calm surface. On nights like this Daubenton's bats can hunt throughout the lake without any problems with clutter from ripples.

Daubenton's bat *Myotis daubentonii* and emerging mosquito, Sweden.

Quite often, a steep sweep is combined with a shallow sweep in one and the same pulse. The search pulses of most bats start on a high frequency before descending in a glissando to flatten out at a lower frequency, while those of a few species ascend instead.

When the prey is detected and therefore close by, the pulses shorten, and the sweeps steepen. Such pulses are called attack pulses. As the bat approaches the insect, the most important task is to estimate the distance. The range of the pulse becomes less important. The pulse repetition rate increases throughout the attack, and just before capture, the bat is transmitting 200 pulses a second which merge into what we perceive as a "buzz". For this reason, such sequences are usually called "feeding buzzes" and indicate that the bat is feeding. Nevertheless, each echo is analysed separately, but the brain uses the echoes to build a sequence revealing how the position and movement of the prey shifts in relation to the bat itself. The muscles of the bat's vocal cords are some of the nimblest amongst animals and can contract roughly a hundred times faster than a typical mammalian muscle.

Clutter and Water Surfaces

Ideally, the only echoes generated in the open air would be those of insects, but in reality, the bat has much more to take into account. Buildings, trees, leaves, the ground, and water surfaces create intrusively strong echoes that easily drown out the weaker echoes from insects, especially if they reach the ears at the same time. Such echoes are referred to as clutter and make echolocation a much more complex tool to use. It is hardly surprising, then, that insects do all they can to conceal themselves amongst clutter in order to avoid discovery.

Many bats minimise the problem simply by hunting in the open air and using long, shallow pulses to maximise their range. Bats that hunt closer to vegetation or at ground level would have serious clutter problems if they used such pulses, so instead they use steep, broad-spectrum sweeps to pick out as many details as possible but at the expense of range. They also keep far enough away from the background that the clutter it causes reaches their ears after the echoes reflected by insects in the air.

Another tried and trusted way of avoiding clutter is to skim the surface of still waters while directing pulses straight ahead so that all the echoes from the water are reflected away from the bat. This makes the water surface invisible and the echoes from insects distinct. This technique is used by Daubenton's bat, amongst others, which flies in straight lines or patterns just above the surface of the water. The surface must be completely calm and open, as vegetation and the slightest wave or ripple create clutter and makes the detection of insects harder. On windy nights, the bat seeks out sheltered waters or moves in over land and hunts amongst the trees.

Doppler Effects and Constant Frequency

Some bats can handle the clutter problem effectively and hunt freely amongst leaves and branches. They exploit the Doppler effect, which is the change in sound frequency caused by movement. Sound is received at a lower or higher frequency, depending on how the emitter moves relative to the receiver. An ambulance siren, for example, sounds higher in frequency as it approaches than when it passes and becomes lower as it disappears again. This principle is used by horseshoe bats and round-leaf bats, which emit long, high-amplitude pulses of constant frequency. When such a pulse strikes a flying insect, the echo returns to the bat at a slightly shifted frequency, caused by the movement of the insect relative to the bat, a discrepancy that the bat uses to calculate its direction and speed—in much the same way as traffic police used radar to catch speeding vehicles before the advent of laser technology. Since the insect's wing beats also create Doppler effects, and given the inverse relationship between size of insect and wing beat rate, the bat is also able to form an impression of the size of its prey.

For the Doppler effect to be really useful, a few specialisations are needed in the bat ear and brain. Horseshoe bats are extremely sensitive

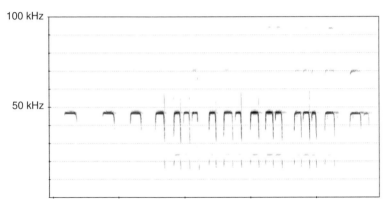

The horseshoe bats can hunt efficiently among leaves and branches because they have an inbuilt filter that separates the clutter from echoes of interest. They use long signals with a single frequency and concentrate the attention to echoes that has been shifted in frequency when reflected from a moving target. Each pulse starts and ends with a sweep, however, which is used to determine the range.

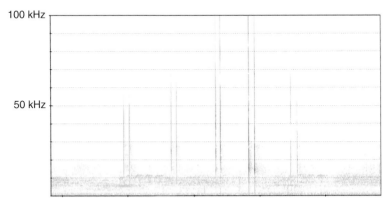

The echolocation calls of the Egyptian fruit-bat consist of short clicks, formed by the tongue against the roof of the mouth. Calls made this way are broadband, including all frequencies at the same time. They are crude and used only to find the way around in the cave roost. As can be seen on the picture, the corner of the mouth remains open. Calls are emitted pairwise, first from one side of the mouth, then from the other.

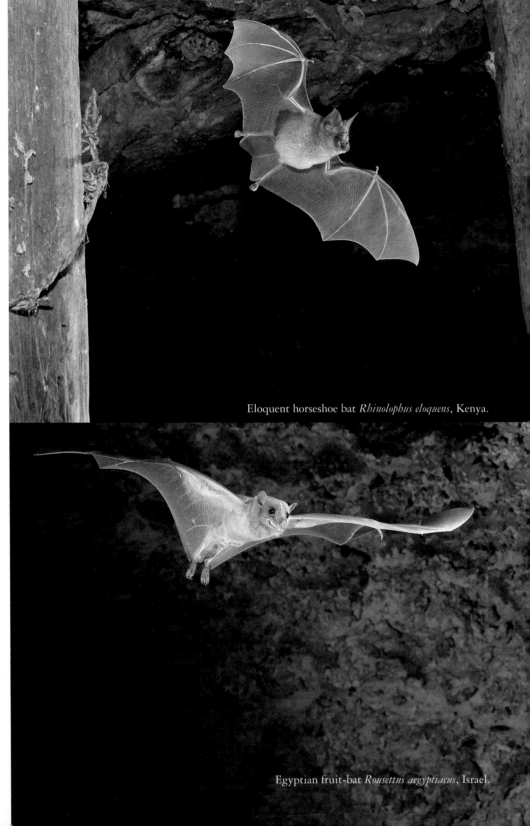

Eloquent horseshoe bat *Rhinolophus eloquens*, Kenya.

Egyptian fruit-bat *Rousettus aegyptiacus*, Israel.

to sounds of a single frequency, which differs between species, and insensitive to the frequencies either side of it. Consequently, they are deaf to their own sounds but not to the Doppler-shifted echoes they receive and can thus call and listen at the same time and use long pulses with no intervening silences. This puts them at a considerable advantage when scanning for flapping insect wings amongst clutter. Unlike other bats, horseshoes can build up an almost continual image of their surroundings, in contrast to the stroboscopic snapshots described above.

The horseshoes and their kin use much higher frequencies than other bats, in extreme cases 150–200 kHz. High frequencies enhance the Doppler effects and presumably make the movements of prey even more distinct. This must be very much in the bat's favour if it can offset the disadvantage of the short range of high-frequency sound. There is still much to learn about the horseshoe bat's echolocation. It is thought, for example, that they are able to create a mnemonic map or sonic blueprint of its surroundings by using the Doppler effects created as it flies along a path or inside a cave.

Interference and Eavesdropping

Just before twilight, bats gather inside the exit hole of their roost, and as the darkness descends, the throng grows denser and denser until it is dark enough for them to evade the predators waiting outside. So, they circle, waiting for the right time. One might wonder how they can decide which echoes to heed and which to ignore or how they can even hear echoes in the racket going on around them. Bats tend to fly in the same direction, which relieves some of the chaos, but it does not explain how they solve the problem of sound interference from their fellows, which they seem to do with ease. The fact is we have no idea, but studies of the bat brain indicate that certain sounds tend to amplify some individual nerve impulses and dampen others so that their own echoes are more prominent and audible than other sounds. So, the ultrasonic bat repellents that people set up outside houses, wind turbines, and so forth to evade bats are unfortunately quite ineffective.

A hunting bat is faced with a completely different situation. Detecting weak echoes from insects requires absolute attention, which is a lot to ask for in a screeching colony of bats. Hunting bats prefer to be alone, but unfortunately for them, this is not always possible. Their echolocation is heard from afar, and an increasing pulse rhythm announces a successful catch—and a bat that has found rich hunting grounds is soon joined by its hungry fellows. Such eavesdropping is the reason why many bats quickly gather around insect swarms.

Apart from the difficulties a bat has concentrating on its own pulses and echoes in the presence of other noisy individuals, it also has to contend with deliberate acts of sabotage. At the very moment a bat homes in on an insect, another can emit a jamming signal and steal its prey from under its nose.

Bats rarely hunt in fog, since tiny airborne water droplets are set in motion by high-frequency sounds which absorbs virtually all the energy rather than reflecting it. For an echolocating bat, a bank of fog, being thus echoless, is like a black hole. So, on foggy nights, bats ascend to clearer altitudes—if they venture out at all. Rain is another matter altogether, and the effect it has on echolocation depends on the size of the water drops. Drizzle is very much like fog, while heavier rain is easier to handle since larger drops do not absorb high-frequency sound to quite the same degree. While raindrops and insects are sometimes hard for bats to tell apart, insects rarely fly in a downpour. Bats also hate to get wet and usually wait out the rain under shelter. A thunderstorm, on the other hand, indicates a change in air pressure which normally galvanise insect—and therefore also bat—activity.

Flying Foxes and Echolocation

Flying foxes, which rely on their visual and olfactory faculties, do not have the problem of using echoes to locate small airborne objects and thus differ from all other bats. Most flying foxes live in colonies in trees and do not use echolocation at all. Some species, however, live in caves and have evolved their own means of navigating in the dark. Some of

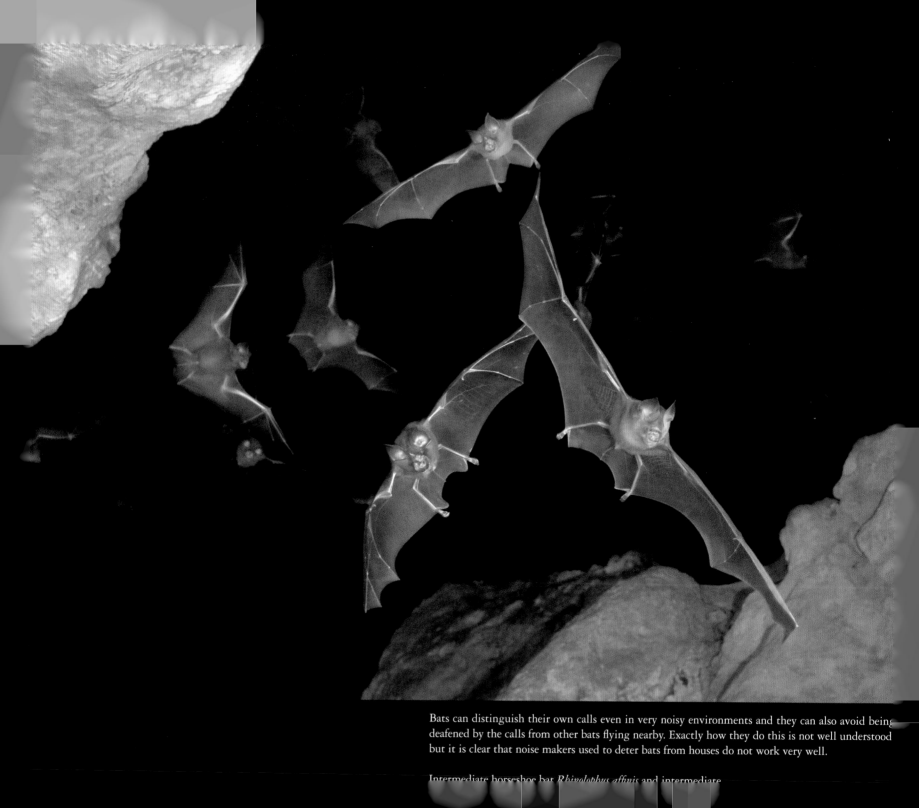

Bats can distinguish their own calls even in very noisy environments and they can also avoid being deafened by the calls from other bats flying nearby. Exactly how they do this is not well understood but it is clear that noise makers used to deter bats from houses do not work very well.

Intermediate horseshoe bat *Rhinolophus affinis* and intermediate

Bats move around protected by darkness. When night falls and they feel secure, they leave the day roosts, guided by silhouettes against the sky. While echolocation is used to scan the immediate surroundings, vision takes over at longer distances, where the echoes die out.

Daubenton's bat *Myotis daubentonii*, Sweden.

Males of some species have scent glands in the corners of the mouth. The glands become obvious in the mating season in late summer and emit a strong and irresistible scent that even humans can enjoy.

Greater noctule *Nyctalus lasiopterus*, Portugal.

them echolocate by using ultrasonic clicks formed by flicking the tongue against the roof of the mouth. These sounds are of a broad spectrum, containing many frequencies emitted simultaneously and thus sound toneless, and are produced in pairs at the side of the mouth—rather than through the front of the mouth—suggesting that they are directed first to one side and then to the other. This suggestion has recently been shown to be correct.

It was recently discovered that some flying fox species in Southeast Asia create clicks by flapping their wings to determine the distance to landing sites and other larger targets. Although this was first written about in the 1980s, it has since been examined more thoroughly and raises the tantalising possibility that more species of flying foxes have developed similar techniques and may not be entirely without echolocating faculties after all.

Vision

Echolocation does not replace the other senses. Bats are not blind, and some species actually have quite good vision. As dusk descends, they turn their gaze to the horizon towards objects silhouetted against the

49

Fringe-lipped bat *Trachops cirrhosus*, Belize.

This large Central American leaf-nosed bat searches for noisy insects such as stridulating crickets or drumming cockroaches but also goes for calling frogs. In any case the bat must be able to distinguish the songs from different species because some are extremely poisonous. Warning coloration does not work at night and it seems likely that the protuberances in the bats' face are sensors that warn for poison before it is too late.

sky. While echolocation is used to scan their immediate environment, vision takes over at greater distances beyond the range of ultrasound. The bat eye has a light-sensitive retina able to register objects in the small amount of light that the night offers. However, its focusing powers are rather crude, and many bats can see nothing smaller than a centimetre-wide object from a metre away, although for the purposes of orientation by shape or the shifting patterns of light and shade in its environment, this is quite adequate. Other species have sharper eyesight and use their eyes as a complement to echolocation while hunting for insects or other prey. Bats can also see ultraviolent light, which enables them to pick out tropical fruits and flowers and, perhaps, to find insects at night. The bat eye acts as a night-vision scope that creates a bright but grainy image. Bats therefore see best in twilight, any lighter, and they are easily dazzled.

A quick look at a flying fox is enough to see that it lives in a different sensory world compared to its echolocating cousins. The flying foxes have no nasal growths, no huge ears, and no elaborate facial adaptations to enhancement of ultrasonic pulses. To our eyes, they look more like regular

mammals, with the large eyes and elongated snouts characteristic of those living in a world of sight and smell. One property that sets them apart, however, from all other members of the animal kingdom is the structure of the choroid, a membrane at the back of the eye that normally provides the retina with nutrients and oxygen. The flying fox choroid has special, capillary-rich projections that feed into the retina, significantly enhancing both the blood supply to the retina and the image formed there.

Hearing

Bat hearing is not confined to ultrasound but can also include lower-frequency sounds, which travel much further and therefore are more suited to communication. As a result, many of their social sounds, such as songs and those used for mating and bonding between females and pups, are fully audible to humans. Some bats are particularly sensitive to sounds in the lower register, a capacity that they use to detect the movements of their prey, picking up clues like buzzing wings, rustling leaves, or insects' footsteps. Long-eared bats are exactly that, with body-length ears that they can swivel to track sounds like parabolic antennas or stretch wide while flying to pick up the faintest of noises, like a moth landing on a leaf or the buzzing of a tiny insect. The advantage of hunting silently like this is that the bats are able to avoid detection by hearing insects, a technique that is actually quite common in the bat world. Natterer's bat hunts houseflies amongst the occupants of cattle sheds, listening for the special buzzing sounds that accompany their mating. While noisy male flies get to mate more, they do so at the risk of their lives.

Taste and Smell

As the sun goes down, the flora of the night awakes and light-coloured flowers unfold and spread their scent. Bats know where to find the flowers and the perfumes they emit guide them through the darkness. For natural reasons, the fruit- and nectar-eating species have the best sense of smell, with noses roughly ten times more sensitive than ours, enabling them not only to distinguish between different species of flower and fruit and be guided long distances by their scent but also to differentiate between different degrees of ripeness. Some insect-eating bats also have a well-developed sense of smell which they use to distinguish between edible insects and those that are either poisonous or foul-tasting.

The sense of smell is used not only to find food but also to communicate. Females in large colonies mark their pups with their own smell for the purposes of familial bonding and of identification in the dark amongst the crowd. The scent glands are also used by males to mark their territory and to attract mates. The glands, which are located at the corners of the mouth, on the chin or on the wings, secrete a characteristic smell that is easily recognisable even for humans. Many bats create a cocktail of this secretion and urine to create a unique, irresistible scent. Some tropical species of the sheath-tailed family store the "perfume" in sac-shaped receptacles in their wings and release it either by flapping the wings or by rubbing themselves against branches and other objects.

Taste and smell are closely related senses, and the bats, like humans, use both to tell the edible from the inedible. The fringe-lipped bat from tropical America hunts for noisy insects like grasshoppers and cockroaches as well as frogs, listening to the direction of the clicks, chirps, or croaks and for any clues as to what is making them. It also must be able to distinguish between the different species' songs, as capturing the wrong prey can prove fatal. Many frogs possess a highly toxic skin, and since the colours and patterns that normally warn off predators are ineffective in the dark, the fringe-lipped bat, apart from its sensitive hearing, is equipped with sensors and projections around the mouth probably used to detect a toxin before it is too late.

Misericordia

In the choir of Hereford Cathedral, near the bishop's seat, are rows of folded chairs "misericords" on which the underside has somewhat strange but beautifully carved figures. The seats also have little shelves used to lean against during the eight different services, which during the Middle Ages could have been performed through the day and night. The participants were expected to stand up during much of this time. Similar carved seats can be found in other cathedrals in Europe. Among the motives are creatures from the so called bestiaries of the time, fantasy monsters such as unicorns and "green men". In some cases there are also bats, which suggests that a bat was considered something else than just an ordinary animal. The bat on the picture was carved around 1380, which means that it is contemporary with the beginning of the great explorations.

"Mappa Mundi", the largest surviving medieval map of its kind is displayed on the wall of the same church. It was made around 1300 and hence depicts the known world nearly 200 years before Columbus. It shows animals and humans found in other countries and also strange creatures believed to exist in yet unexplored corners of the world.

Hereford Cathedral, England.

Mosquitoes (Culicidae).

Crane flies (Tipulidae).

House-flies (Muscidae).

Dance flies (Empididae).

Window-gnats (Anisopodidae).

Fungus gnats (Mycetophilidae).

Crane flies (Limoniidae).

Moth flies (Psychodidae).

Flesh-flies (Sarcophagidae).

Winter gnats (Trichoceridae).

Hunting and Feeding

BATS EAT THE equivalent of half of their body weight every night. For insect eaters this can mean more than a thousand small insects, which in terms of energy is roughly the same as a person eating 30–40 hamburgers. Suckling females need an especially high-energy intake and can eat even more. So as not to put on excess weight, bats are careful to only eat what gives them most energy. Fruit eaters spit out fibrous matter, stones, and pips, and insect eaters usually bite off the indigestible legs and wings of their larger prey. The abdomens of swarming insects like ants and termites are rich in fat, and in order to maximise their nutrient intake, bats sometimes feast only on that particular part and spit out the rest. Many bats eat on the wing, while others, particularly those that hunt larger animals, prefer to land before savouring their meal in peace and quiet. Bats chew their food carefully to ease digestion and, owing to the large size of their stomachs in relation to their intestines, can eat a great deal when the occasion arises. At the same time, bats can digest their food and expel the waste in less than an hour.

Insects

Most bats live on insects and have probably always done so. Their prey comprises everything from tiny midges and aphids to the largest beetles. Many are very flexible in their choice of diet and eat whatever is available.

The dipterans or flies are very important as bat food. They often occur in large numbers almost everywhere and many species replace each other throughout the season. Most species breed in water, which is why lakes and rivers are important as feeding sites for bats.

The dipterans or flies are very important as bat food. They often occur in large numbers almost everywhere, and many species replace each other throughout the season. Most species breed in water, which is why lakes and rivers are important as feeding sites for bats.

Each species of insect swarms for a short time, usually for a few days or a week, but with the thousands of insect species available, there is always food to be had throughout the summer. Insects also swarm at specific places and in specific environments. Bats have learnt where and when the insects are available and are quick to make the best of the occasion. Insects that emerge from water, such as chironomids—nonbiting midges— and caddisflies, are particularly important sources of nutrition in the Nordic region, since there is a great abundance and a continual turnover of species throughout the summer. Some bats are, of course, better at exploiting certain sources of food than others. Large bats, on average, catch larger insects than smaller bats, but the variation is wide and the exceptions many. Even more than this, the nature of the prey depends on the hunting technique applied by the hunter. Brown long-eared bats and barbastelles are moth specialists, serotine bats and northern bats prefer dung beetles, and the soprano and Nathusius' pipistrelles mostly enjoy a diet of flies such as mosquitoes, midges, crane flies, and gnats.

Aeroplankton

As we have already mentioned, swift narrow-winged bats often hunt at high altitudes, elegantly mastering the dark with acrobatic prowess as they hunt for swarms of tiny insects. Low-frequency, high-amplitude echolocation, as used by these bats, has a comparatively long range, and every

© Springer International Publishing AG, part of Springer Nature 2017
J. Eklöf, J. Rydell, *Bats*, https://doi.org/10.1007/978-3-319-66538-2_4

55

echoing response reveals the whereabouts of the prey. The bats systematically scan their surroundings, picking out landmarks on the horizon with their eyes. The species of bent-wing bats stick together in large colonies that flock to wherever insects are hatching. These bats are relatively small and light but quick on the wing. They feed partly on aeroplankton carried by air currents and winds high above the ground. Aeroplankton includes animal life forms, such as midges and aphids as well as spiders and caterpillars that sail through the air suspended from their gossamer threads. Colonies of bent-winged bats are often enormous, containing tens of thousands on individuals, and, like bats in general, are important for agriculture and other activities as a natural form of vermin control.

Above Water

Watercourses and wetlands are the most important hunting grounds for bats in the north, as mangrove swamps and estuaries are for many tropical and subtropical species. Water is the environment for a wide variety of insects in the larval stage, such as mayflies, caddisflies, and most true flies including mosquitoes, gnats, and nonbiting midges. After hatching, some of these insects congregate in swarms above the surface of the water or above trees along its edge. Flies are easier for bats to catch than many other insects since they are earless and therefore deaf to echolocation. Most true flies are active for a brief period at dusk after the birds have returned to their night roosts but before the bats have started to venture out, but they also find protection by forming dense swarms based on the same safety-in-number principle as a shoal of fish or a flock of starlings.

Even though many bats hunt over the water, few of them catch fish. This is because it is not easy to locate prey under water in the dark. There might have been the odd report on Daubenton's bats being able to catch fish, but it is in Latin America where the real specialist, the fishing bat, lives. The fishing bat sweeps across the dark surfaces of rivers, trawling with its long hind legs, large feet, and hook-like claws. The fishing bat has a characteristic echolocation technique and can sense the slightest ripple on the surface caused by a fish.

Fishing bats, Daubenton's bats, and many other over-water hunters have large feet for trawling, a technique so named because it was once thought, erroneously as it turns out, that they drag their feet along the water in a more or less random fashion. But as we can clearly see in the photograph overleaf, a bat is perfectly able to locate its prey with pinpoint precision using echolocation alone and pluck it from the surface without any recourse to chance.

Drinking

In hot, dry climates, bats need to keep themselves hydrated, especially suckling females, which therefore need to drink every day. There are some desert species that can live off the water content of the insects they eat, but many of the inhabitants of hot, arid regions are geographically constrained by the need to drink. In some areas, such as the Mongolian steppes, bats are totally dependent on human wells, and as rural communities decline, the wells collapse, and the bats disappear. Bats also visit dug wells and livestock troughs in many other countries, and in California, southern Europe, and other hot places, it is not unusual to find bats quenching their thirst from outdoor pools.

Bats prefer to drink on the wing from still water surfaces. To avoid getting wet, they need to perform a precision manoeuvre, but given that still waters return no echo to an individual approaching from the side, this is difficult to do. Therefore, bats approaching a watering hole to drink always circle a few times to pinpoint the surface exactly before descending.

Hanging and Waiting

Instead of hunting on the wing, certain bats listen for their prey while hanging silently from a branch. Yellow-winged bats hang in the open tropical air, fully exposed to predators, from the branches of the acacia tree. But its colour serves as camouflage, its large eyes and ears indicate good vision and excellent hearing, and it is constantly on the alert. Yellow-winged bats live in pairs and engage in territorial defence,

Least bent-winged bat *Miniopterus minor*, Kenya.

At high speed on long narrow wings, the bent-winged bats search for small dipterans, aphids, spiders, moth larvae and other aerial plankton at high altitudes. The colonies usually consist of thousands of individuals and throughout the warmer parts of the globe they constitute an important resource for humans as natural pest control agents.

African bent-winged bat *Miniopterus africanus*, Kenya.

In the middle of Darwin, Australia's northern metropolis, a set of drain pipes empties into Rapid Creek, which carries the rain water to the mangroves at the coast, where it meets the tide. In the pipes there are bats of several species, including an Australian relative to Europe's Daubenton's bat.

Large-footed myotis *Myotis macropus*, Australia.

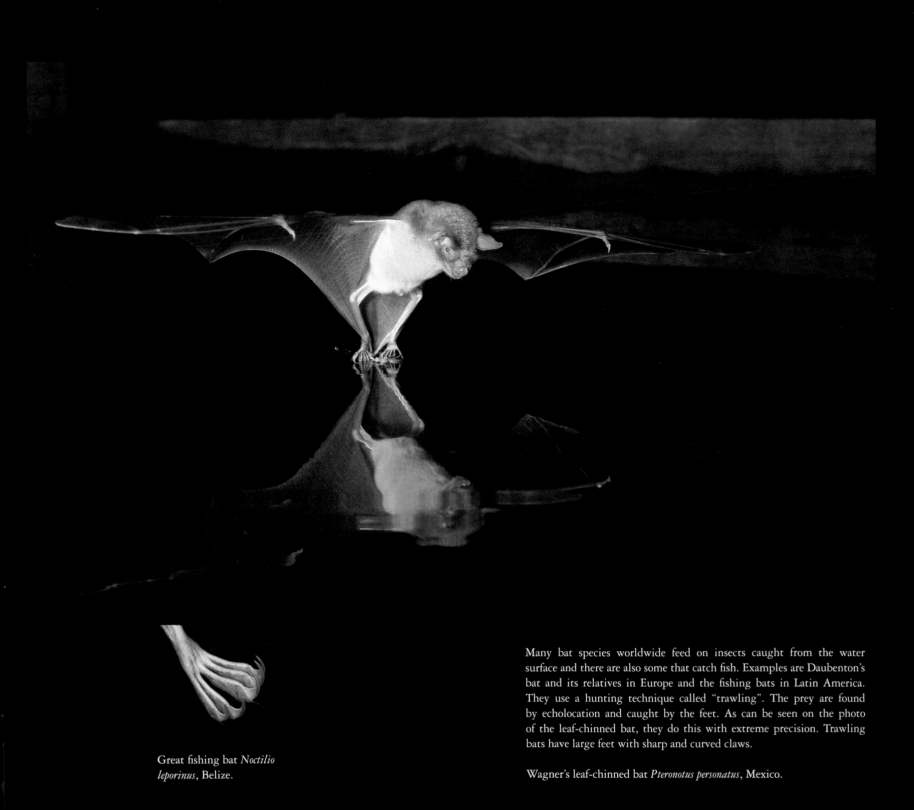

Many bat species worldwide feed on insects caught from the water surface and there are also some that catch fish. Examples are Daubenton's bat and its relatives in Europe and the fishing bats in Latin America. They use a hunting technique called "trawling". The prey are found by echolocation and caught by the feet. As can be seen on the photo of the leaf-chinned bat, they do this with extreme precision. Trawling bats have large feet with sharp and curved claws.

Great fishing bat *Noctilio leporinus*, Belize.

Wagner's leaf-chinned bat *Pteronotus personatus*, Mexico.

sometimes with their adult young, which is an unusual lifestyle for bats. The female and male always keep company for a while in the afternoon, grooming their coats ahead of the evening's hunt. As twilight approaches, both bats hang from a branch and wait, heads swinging back and forth and eyes registering the slightest movement in the dying light. Directing their ears independently, they listen for echoes and, perhaps even more importantly, for sounds generated by their prey, such as wing beats and rustling in the undergrowth. Then targeting their prey, they swoop, snatch it up, and return to the branch.

This hunting technique of patient waiting and sudden pouncing is thought to be a primitive strategy adopted by early bats, maybe even before they were fully developed fliers. It is roughly the same technique as used by the flycatchers of the bird world and, occasionally, by modern-day slit-faced bats and false vampire bats. Even the larger species of horseshoe bats hang in wait like this, although they echolocate with their monotone pulses instead of passively listening for sounds. Sitting still waiting for prey is more energy saving than flying, so the technique suits the heaviest species and times when food is sparse.

Gleaning

Not all bats hunt flying insects. Some pick them straight off of leaves and branches or from the ground, a hunting technique called gleaning. This requires a different set of sensory skills and flying technique, as the echo reflected by the insect must be distinguishable from that of the background of a leaf or blade of grass. Bats that hunt close to vegetation therefore use broad-range, steep sweeps that generate echoes packed with information. They also take into account other sounds they hear, including those generated by the prey, as well as olfactory and visual clues. Their wings are short and broad, and the ability to hover in the air and manoeuvre in confined spaces takes priority over speed.

Some species hunt close to the ground. The greater mouse-eared bat, for example, hunts for beetles along the forest floors of Central and southern Europe. It falls completely silent as it approaches the leaves on the ground, where it hovers listening for the slightest rustle and analysing the different smells with its sensitive nose. Swivelling its ears to pick up the footfalls of a ground beetle scuttling through the undergrowth or a moth launching itself from the grass, the bat sweeps over the ground with its tail membrane formed into a scoop that snatches up its prey in the blink of an eye. The brown long-eared bat also prefers hunting by stealth so as not to alert hearing animals to its presence. Its ears, which are amongst the most sensitive in animals, listen for the flapping of wings or the scrape of feet against leaves, while its eyesight is sharp enough to pick out even rather small insects.

Some large bats in warmer climes occasionally catch larger prey, such as scorpions, reptiles, rodents, and sometimes even other bats using either a similar stealth-based hunting strategy or the more energy-efficient hang-and-wait technique.

Changing Strategy

Brandt's bat sleeps with spiders' webs caught in its fur in the autumn, a rather conspicuous clue as to what and how it hunts. Normally, it flies in straight lines along forest paths in the northern taiga, around crofts and shacks, catching insects where no other bats bother to look. Come the autumn, as the nights grow colder and insects scarcer, the ground is carpeted by the web of the sheet web spider, which hangs underneath to protect itself from the weather and – evidently with little success – hungry Brandt's bats. Including spiders in its diet probably explains why this species thrives so well in northern climes, where the insect summer is actually far too short for bats.

It is not just Brandt's bat that changes strategy when the summer-time insect population starts to decline. In fact, it is quite common for bats to adapt to the prevailing conditions. In southern Europe,

To drink from a calm surface in the dark is not trivial. For a bat closing in at the pond from the side, there is a risk that the echoes from the surface reflect away and leave what may look like a dark hole. Without echoes the bat may risk ending up in the water. Fortunately bats are good swimmers.

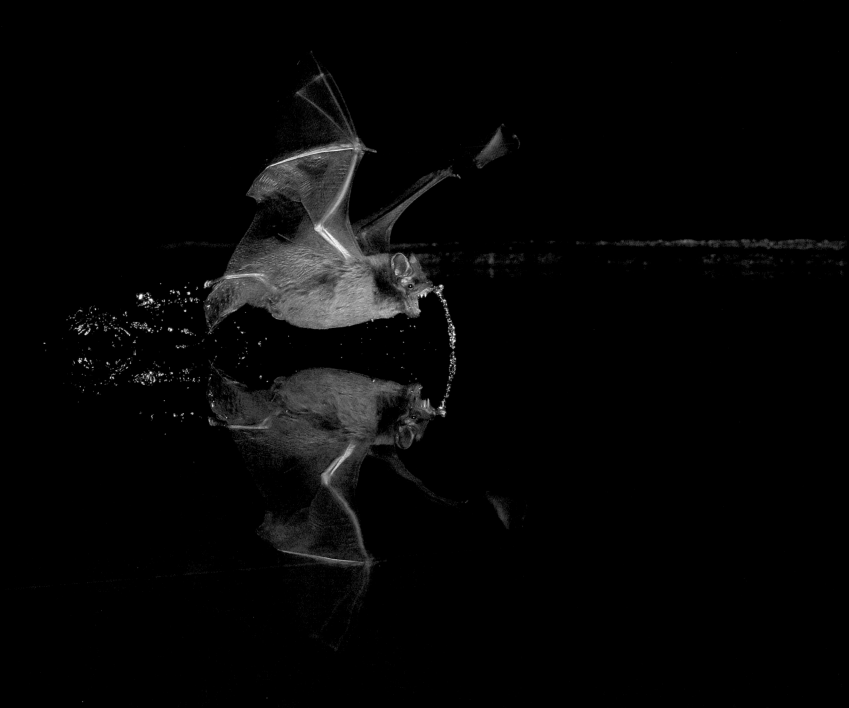

Lesser noctule *Nyctalus leisleri*, Italy.

Among the ruins of the old Arabic town of Gede at the coast of East Africa lives a magnificent bat, a giant among the insect eating species. Weighing about 100 grams it is too heavy and not quick enough to chase insects in the air. Instead the bat listens for the prey hanging from a branch.

Striped roundleaf bat *Hipposideros vittatus*, Kenya.

Brown long-eared bats sometimes live in bird boxes and in this case the bat colony has occupied the same box for more than 10 years. The long-ears pick prey from branches and leaves, as well as from walls and roof spaces. Using a special hunting technique and extraordinary hearing ability they discover and capture prey animals that move about in cracks and crevices or in their webs, those that few other bats can find.

Brown long-eared bat *Plecotus auritus*, Sweden.

In the old Maya-ruins lives this large leaf-nosed bat. It feeds on vertebrates including other bats and forest birds. Feathers and other remains piled up inside the bat's roost tell the story.

Woolly false vampire bat *Chrotopterus auritus*, Belize.

Brandt's bat or the taiga bat lives in the northern coniferous forests belt from Scandinavia to the Russian Far East. It is an expert survivor in many ways. In the autumn, when nights get colder and insects become harder to find, the ground in the taiga is still covered in spider webs, each with a spider sitting underneath. The web forms a roof and provides protection against the weather and perhaps also against foraging taiga bats.

Brandt's bat *Myotis brandtii* and spider *Linypha hortensis*, Sweden.

Springtime evening and the bat is leaving the box in Coto Doñana, aiming at the millions of migrating birds that arrive to southern Europe from Africa. This species is studied intensively in Coto Doñana. The ring on its forearm is a permanent individual mark that contributes data about its survival and movements. The hole in the wing membrane is after a skin punch taken to obtain a DNA sample that will provide information about the relationship with other individuals and also about virus infections. It heals in about a week. The cable visible in the box serves a mini-computer that reads a transponder on the bat's back. The transponder reveals where the bat resides each day without the need to catch it.

Greater noctule *Nyctalus lasiopterus*, Spain.

The vampires have become very common in Latin America, particularly in rural areas with cows, horses and pigs providing almost unlimited amounts of food. Vampires were formerly a big problem for farmers, since they spread rabies between the farm animals. Nowadays many farm animals are vaccinated and the problem has diminished substantially. In northern Belize, where this picture was taken, vampires are not of great concern, because rabies is rare or absent.

Common vampire *Desmodus rotundus*, Belize.

Little red flying fox *Pteropus scapulatus*, Australia

just as the migratory birds are heading south, the greater noctules leave the deciduous forests to make for the river valleys of the Iberian Peninsula, where they rise to high altitudes on the winds and currents to wait for passing flocks of small birds that pass. Given their deafness to ultrasound, birds present relatively easy prey for those that can find them in the dark and that are big and strong enough to catch them. The birds are eaten on the wing, and only gently descending feathers tell of the drama playing out high above. For much of the year, when few if any migrating birds are there for the catching, the bats live on insects, just like their smaller relatives, but when the spring arrives and the birds migrate back north, they resume their high-altitude

patrol. That bats have a seasonal diet of migratory birds is a relatively new discovery, although recently several similar observations have been reported from different parts of the globe.

Blood

The real vampire bats live in caves and hollow trees in Latin America. There are three species, and all of them live on blood, making them unique amongst mammals. Two of the species usually drink the blood of birds, while the third subsists on mammalian blood drawn mainly from domestic animals such as cows, pigs, or horses, which the bats can fly several kilometres to find. In Latin America, people often sleep with their windows open, and although it is rare, humans can be fed off as well. Vampire bats use all their senses on the hunt for prey and even use an extra heat-receptor sense, the only mammals to do so. These receptors, which sit around the nose leaf, help the vampire to find a suitable place in which to bite. When the prey is located, it lands on the ground a little way off and covers the last few metres hopping along on its feet. Drawn to the animal's smell and its body heat, it bites a 3–4 mm hole in the skin with its razor-sharp front teeth. Normally, its victim notices nothing and continues sleeping, which is fortunate for the bat as it can take almost 20 min for it to ingest the two tablespoons of blood that its stomach can contain. To thin the blood and promote flow, vampire bats have a special substance in their saliva called draculin, which now forms the active ingredient of a drug designed to dissolve blood clots.

Blood consists mainly of water, and almost directly after settling down to drink, a vampire bat starts to urinate, ridding itself of unnecessary weight and maximising its nutrient intake. After feeding it pushes off with its thumbs and feet and vanishes silently into the night back to its colony, where it shares its food with its less successful

In tropical Australia the large flying foxes leave the day roosts just before sunset. They leave the coastal mangroves and aim towards mango-farms and flowering ornamentals in parks and gardens. This is sometimes to the annoyance of farmers, as they eat a lot of fruit each night.

Black flying fox *Pteropus alecto* and little red flying fox *Pteropus scapulatus*, Australia.

Two rows of flowers open at sunset. Before dawn they have withered and fallen of, emptied of nectar. The flowers fit a long bat-tongue. They are light, in contrast to everything else nearby, and can be seen even in the dark tropical night and the open leaves also provide an acoustical guide, just in case. Modern bananas are propagated by shoots, but the nectar production continues as a remainder of former times.

Pallas's long-tongued bat *Glossophaga soricina* and banana *Musa acuminata*, Belize.

There are hundreds of species of wild peppers in the tropics. Many of them are bushes growing in disturbed areas, where the original forest has been cut. They grow fast and soon cover the open areas. The fruits are located on long inflorescences, not so obvious for human eyes but shaped to attract bats. Bats of the genus *Carollia* are specialists on pepper fruits, which are not left alone for long after they are ripe. The seeds are spread along tracks and roads by flying bats.

Seba's short tailed bat *Carollia perspicillata* and wild pepper *Piper* sp., Belize.

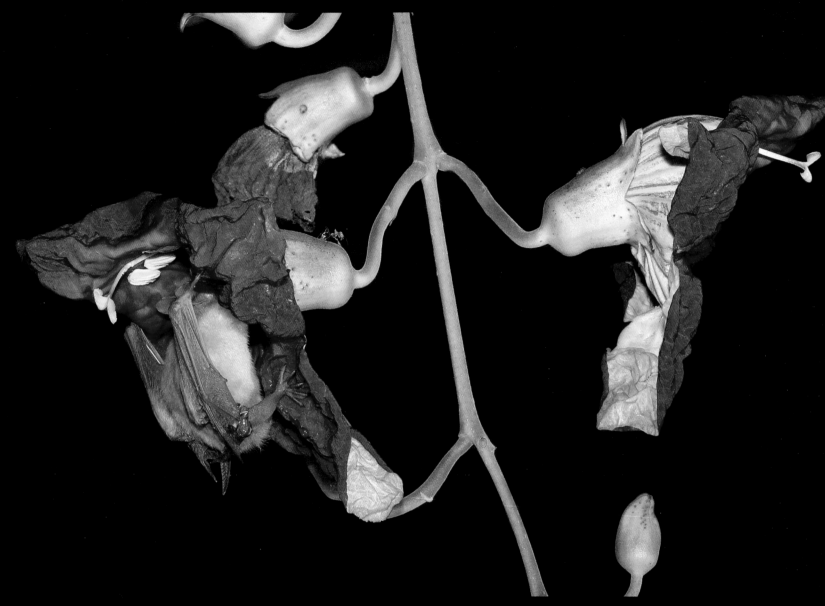

The bud unfolds in the evening exposing the flowers. Dozens, sometimes hundreds of flowers spread their characteristic scent. But already before dawn they have fallen to the ground. The sausage tree originates from Africa, where it is visited by large flying foxes searching for nectar. But it has become established as an ornamental tree in many tropical countries and in Australia a minute flying fox stands on its head in one of the oversized flowers. This bat is a dwarf among flying foxes, weighing in at only 10 grams.

Lesser long-tongued nectar-bat *Macroglossus minimus*
and sausage tree *Kigelia africana*, Australia.

fellows. Since vampire bats can live no longer than a couple of days without blood, such altruism is vital to the survival of the colony, and the individuals that share their food can expect to have their generosity rewarded the next time they are in need.

Fruit and Nectar

Two bat families, flying foxes of the Old World and the leaf-nosed bats in the Americas, have developed an appetite for fruit and nectar. Flying foxes are strict vegetarians, and when not eating fruit and nectar, some species also feed on leaves, despite lacking the enzymes needed to break down cellulose. They simply suck out the juice and spit out the indigestible residue. Pollen also provides an important source of protein for these species. Some leaf-nosed bats complement their otherwise vegetarian diet with the odd insect, perhaps for precisely its protein content. Conversely, some insect-eating bats complement their diet with fruit and pollen.

Unlike insects and other prey, flowers and fruits are immobile, which makes them harder to echolocate. Bats therefore find it easier to find fruits and seed pods that stick out conspicuously from branches and trunks. Many tropical rainforest plants seem to be adapted to bats, which pollinate their flowers or spread their seeds. Some plants have parabolic structures, such as concave leaves or flowers, which return stronger echoes and show bats exactly where fruits or nectar are located. Otherwise, fruit- and nectar-eating bats are led by their visual and olfactory senses. Bat-pollinated flowers are often either light-coloured or red with bright, contrasting patterns and usually have a powerful, characteristic scent.

The great diversity of tropical fig trees sustains fruit eaters, such as apes, birds, and bats, with their abundant produce. They come into flower with irregularity at any time of the year, but when they do, they do so with gusto. Fig trees often start life up in another tree, perhaps in the fork of a branch, where a seed has been deposited by a bird or a bat. Their aerial roots grow into trunks and eventually embrace their host,

into which they take root. Several Latin American species of leaf-nosed bats have a predilection for figs, as often evidenced by their powerful jaws. Outgrowths around the mouth contain sensory cells that are probably used to judge the quality and ripeness of the fruit.

Leaf-nosed bats rarely stay to consume their meal in the tree, preferring to take it to where they can suck out the juice and eat the flesh in peace and quiet. The fibrous material is spit out with the seeds, which are thus spread throughout the forest.

Every night, flowers open to reveal their troves of nectar. Nectar-eating bats are guided by the scent and return evening after evening to the same spot, hovering over the flowers, which are visible to them even in the dark tropical nights. These bats have long snouts, small teeth, and sometimes a tongue longer than half their body, with which they probe deeply into the inflorescences. The tongues are unusual in being covered with hair-like structures that, when engorged with blood, become erect and help the bat lap up the nectar.

Seeding and Pollinating

Over 500 disparate flowers around the planet, including many economically important species such as agave, eucalyptus, balsa, and guava, are more or less dependent on bats for their pollination. In Southeast Asia, nectar-eating bats are essential to the propagation of durian trees, the fruits of which are multibillion industries in Thailand and Malaysia. Flying foxes and leaf-nosed bats also plant the seeds, spilling fruit residues and excrement wherever they go, and the forest grows in their wake. The fig-eating bats of South and Central America can easily fly 10 km to find ripe fruit and along the way, deposit a rain of seeds that establish themselves and take root. In clearings and along the verges of roads, the bats sow new trees, giving the rainforest a chance of regrowth.

While flying foxes are similarly invaluable to the African ecosystems, they are considered pests by farmers, who hunt and harass them. They are sometimes sold as bush meat, and a deeply rooted superstition that associates them with evil spirits merely exacerbates their persecution.

Eudonia sp.

Bilberry emerald *Jodis putata*.

Twenty-plume moth *Alucita hexadactyla*.

December moth *Poecilocampa populi*.

Arms Race

OVER THE MILLIONS of years that bats have evolved increasingly sophisticated hunting techniques, their prey species have evolved ever more effective means of defence. In fact, as they have more to lose, their progress has been faster. To the insects, it is a matter of life and death while to the bats one "merely" of getting a meal.

Insect Ears

Many nocturnal insects have well-developed hearing, which is their first line of defence against bats. Moths, grasshoppers, crickets, praying mantises, and green lacewings are just some of the insects with ultrasonic hearing. The ears of the different insect groups have different structures and are located on different parts of the body, suggesting that they evolved independently at different points in time. In moths alone there are at least nine distinct types of ear. Most or perhaps all insect ears seem to have evolved in response to bats, the prevalence of this adaptation indicating that bats have had a profound effect on insect evolution.

Because of their powers of hearing, these insects have become harder to catch as they can hear echolocation signals early enough for them to reach

There are many insects that can hear ultrasound, including some nocturnal moths. With ears they can detect the enemy far away and this works as an early warning device. Thanks to this they can fly slowly and maintain a low body temperature, thus saving energy that otherwise would be used in fast, evasive flights. However, there are also deaf moths, but they need to be fast flyers, which in turn requires a higher body temperature and good insulation, and thus waste energy to a much higher extent.

safety. Depending on how close the bat is judged to be, a moth, for example, will either simply change direction or dive to the ground, where it can sit until the threat passes. If a moth hears an approaching bat high up in the air, it performs a loop or a spiralling dive to throw the bat off. Moths preparing to fly immediately seize up when they hear a bat. If they are sitting in a tree, they simply lose their grip and tumble to the ground.

The development of the moth's auditory sense has affected the life and behaviour of these insects in other ways too. Hearing moths often fly peacefully over open terrain knowing that they will hear any predator approaching at a distance. Since they fly slowly, such moths can maintain much lower body temperatures than their deaf relatives and can thus conserve their energy for other things, such as reproduction.

As is well-known, crickets and grasshoppers use hearing as a means of sexual communication. But while the chirping of the males tells the females where they are, it also puts them in peril in a world of hungry bats.

Bats have also made an impact on the evolution of diurnal butterflies, since being active during the day instead of at night is one good way of defending yourself from bats. Butterflies of the Nymphalidae family possess a structure at the base of their wings called Vogel's organ, which has been known to scientists for a century. However, it was not until recently it was recognised as a regressive organ of hearing. In some tropical Nymphalidae species, such as the crepuscular owl butterfly of Central America, Vogel's organ still serves as an ultrasonic detector, helping the butterfly to avoid bats at twilight when it is most active. The Nymphalidae hearing organ suggests that their ancestors

© Springer International Publishing AG, part of Springer Nature 2017
J. Eklöf, J. Rydell, *Bats*, https://doi.org/10.1007/978-3-319-66538-2_5

In potato cellars bats as well as insects can sometimes be found hibernating in winter.

Norhern bat *Eptesicus nilssonii*, fungus gnats (Mycetophilidae) and herald moth *Scoliopteryx libatrix*, Estonia.

Tiger moth, Taiwan.

The bat defence in the tiger moths consists of ultrasonic ears in combination with poison. Their distastefulness is announced by ultrasonic warning signals from a sound organ on the thorax as an approaching bat is detected.

were nocturnal and were hunted by bats—as, possibly, were other diurnal butterflies. However, some Nymphalidae species, such as the northern European satyrines, have become insensitive to really high frequencies and no longer react to bats although their ears do still alert them to other predators rustling towards them in the foliage.

Warning Signals and Mimicry

In some species of butterfly, the defence mechanism has been taken one step further. The peacock butterfly and its relative the small tortoiseshell have partially functional organs of hearing in their wings, which they use in the autumn and winter when hibernating alongside bats. These butterflies are mildly toxic and, when their hearing is triggered by echolocation signals, can produce a clicking sound with their wing

Owl butterfly *Caligo eurilochus*, Belize.

The owl butterflies are active at dusk and possess an efficient defence against bats. It has ultrasonic ears in the ribs at the base of the wings by which it can detect an approaching bat.

muscles that is presumed to deter bats. However, the system is not fail-safe, and piles of bitten-off wings can often be seen littering the floors of the steeples and earth cellars where bats roost.

The members of the tiger moth family (Arctiidae) are usually better at evading bats. Many of them are more or less poisonous, which their gaudy wing patterns signal to birds, and also possess a sound-generating tymbal organ on their thorax with which they create a series of warning clicks whenever an approaching bat is heard. The click trains serve as a warning and have the same purpose as their contrasting wing colours during the day. In other tiger moth species, the clicks can confuse attacking bats by acting as a biological jammer that sends out a smoke screen of false echoes to conceal their exact position.

There are, of course, also instances of insects that use mimicry, bluffing their way to survival by imitating the appearance or behaviour of a completely different species. The warning clicks of the tiger moths, for example, are mimicked by other, non-poisonous moths.

Other Defence Mechanisms

Most insects, like mosquitoes, nonbiting midges, caddisflies, mayflies, aphids, and beetles, are deaf and therefore easy for echolocating bats to approach. But they are not entirely defenceless, even though their strategies have not been as thoroughly studied as the moth sense of hearing. One might wonder, for instance, why fireflies light up the tropical nights. Is their glow simply an interspecies signal or a warning to bats that they are inedible? Fireflies smell and taste bad, at least to humans. One might also wonder why mosquitoes are most active for a brief period at sunset just after the birds have stopped hunting but before most bats leave their roosts. Or why many nocturnal flies, such as mosquitoes, crane flies, and midges, always fly in dense swarms just like bats do when threatened by predatory birds. It is very likely that bats have very much influenced the evolution of such behaviours.

Deaf moths normally fly close to the ground or in amongst foliage, where most bats inevitably have problems with clutter. There are other insects that do the same. The aquatic water strider, which spread-eagles itself on the surface of calm pools during the day retreats to the banks at dusk, hiding close to the land amongst clutter-generating stones and vegetation from hungry Daubenton's bats hunting close by.

The Bats' Response

In order to catch hearing insects, some bats hunt in silence and instead of echolocation make use of their ears to listen out for telltale sounds, like the flapping of wings or the rustling of leaves. So, bats, like many other predators, are quite capable of hunting by stealth, a technique that can also reveal prey hiding in clutter.

Some bats can also use their eyesight to find prey that is attempting to hide. Leaf-nosed bats living in the North American desert hunt scorpions using their vision, at least on clear starry or moonlit nights. In the meadows of Scandinavia, mating ghost moths, which are deaf, dance close to the grass in the hope of avoiding the northern bats patrolling overhead. However, it has been shown that these bats can see the white insects and locate them amongst the clutter of the grass. Every year at the same time, the bats return to the meadows where the moths gather, but as the open grassy lands the moths inhabit gradually disappear, such spectacles are becoming increasingly rare.

The occasions when bats can hunt with their eyes are presumably quite rare, and not all bats are able to completely abandon echolocation to rely solely on the sounds produced by their prey. Some bats have therefore adopted another strategy for discovering hearing insects—they have pushed the frequency of their pulses outside the auditory range of their prey. Some large, swift-flying bats use frequencies of 10–13 kHz, which are easily audible to humans but not to the moths

Long-eared bats usually hunt in silence, listening for sounds generated by prey animals, but may sometimes use low-frequency (10-15 kHz) echolocation calls which are nearly inaudible to moths. The bats' long ears are suited for relatively low frequency sounds, the ears and the sound waves are of similar length, about 3 cm.

Brown long-eared bats *Plecotus auritus*, Sweden.

Some bats use echolocation calls with frequencies so high that they fall outside the range that moths can hear. The small horseshoe bats that live in the old pump station in Coto Doñana echolocate at 115 kHz and therefore can approach their prey undetected.

Lesser horseshoe bat *Rhinolophus hipposideros*, Spain.

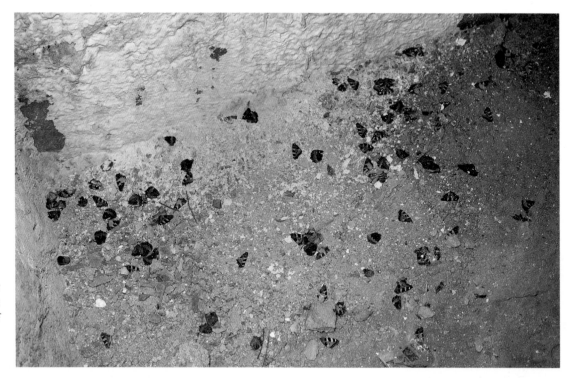

Some insects are brave or stupid enough to share winter roost with bats. It sometimes works but more often not. Piles of butterfly wings indicate the location of the bat's feeding perch.

Tortoiseshell butterflies *Aglais urticae*, Sweden.

they hunt. As we have seen, such frequencies render the bats "blind" to smaller insects but greatly increase their range.

The spotted bat in America uses an even lower hunting frequency of 9 kHz. This species is also an expert moth hunter. Brown long-eared bats can also be heard emitting their loud, low-frequency squeaks in the spring, before the trees come fully into leaf. This is likely an adaptation to hunting moths in open terrain before the foliage starts producing really interfering clutter. During the summer, these sounds grow less common as the long-eared bats start hunting using high-frequency sounds or without echolocation. The low-frequency sound of the long-eared bat is thought to be a communication signal, but it is possible that it performs other functions as well.

Sweden is home to another specialised moth eater, the barbastelle. This bat echolocates at a lower amplitude than most other moth eaters and can thus come as close to its prey as 3 m before being spotted—the same moth would have detected any other bat, such as a northern bat, from ten times that distance. The disadvantage of reducing the amplitude is, of course, that it greatly limits the bat's range and shortens the time it has to react.

In more recent times, bats have discovered a new rich hunting ground for hearing insects, namely, under the glow of streetlamps. Moths only react to ultrasound during the hours of darkness since that is when bats pose a threat. The light from the streetlamps disrupts their normal reactions, leaving the insects easy targets for hungry bats. We will be returning to the effect that streetlamps have had on both bats and insects in a later chapter.

Reproduction

ALL ANIMALS NEED to produce young when food is at its most abundant, and many bats must contend with a source of nutrients that fluctuate widely from one season to the next. In northern latitudes, there is hardly any food at all during the colder half of the year, while the warmer offers an excess. The fluctuation is not as pronounced in the tropics, but there can still be great differences between the dry and rainy seasons, so there too it is important that bats reproduce at the right time.

Nurseries

As spring turns to summer, pregnant females congregate in separate maternity colonies, sometimes returning year after year, for decades, to the same secure place where they themselves were once born. Many of them are related, and they help each other to raise their pups, sharing their experiences and, in some species at least, babysitting for those that need it. The males, meanwhile, remain in their winter roost or live in their own holes to fly around in isolation with an eye on the hatching cycle of their prey. Only in a few species do the males form part of the nursery colony. In most cases they are expelled as soon as they reach sexual maturity.

The bat gestation period is roughly 2 months, but if the weather turns and food becomes scarce, it can be extended by several weeks,

Under the eaves of the boat house resides a little colony of proboscis bats. Each female hides a young, which soon is ready to follow the mothers on the insect hunts along the river. The bats remain motionless, trusting the camouflage makes them invisible, which is usually the case – on a tree trunk.

during which time both mother and foetus enter a state of hibernation in anticipation of better times. This means of fending off starvation is unique to bats and an adaptation to the unreliable availability of food. When bats give birth, they do so with exceptional synchronisation.

The spring temperatures and the first insects stimulate ovulation, and all females in the colony give birth to their young within the space of a few days. In northern Europe, this occurs at around midsummer, just before the insect population reaches its peak, and in southern Europe, 1 or 2 months earlier.

In tropical and subtropical zones, insect emergence and bat reproduction are controlled more by the rainy seasons than by the summer. As the first rain clouds herald the monsoon, the females ready themselves to give birth so that their pups can grow up with a surplus of food during the most abundant time of the year.

In the days prior to giving birth, the females usually remain in the nursery as they are too heavy and ungainly to go out hunting. After giving birth, the young will need all their attention. So, for a few days around the time of birth, the females live off stored body fat alone.

Just before giving birth, the female rotates and hangs, for once, upright. The pup (for there is rarely more than one) is often born feet first, which prevents its wings becoming stuck on the way out. Neonate pups are relatively large and weigh just over a quarter of the mother's own body weight, but the highly elastic ligaments of the female's pubic bone ease their delivery. Flying foxes in captivity have been observed assisting each other when giving birth, with older females instructing the younger ones not to hang upside down.

© Springer International Publishing AG, part of Springer Nature 2017
J. Eklöf, J. Rydell, *Bats*, https://doi.org/10.1007/978-3-319-66538-2_6

A church tower is winter home for a parti-coloured. It is dry and dusty and sometimes windy, but peaceful. Male bats usually live alone and to sleep next to the organ and the church bells does not seem to be a problem. The bells and the music toll at frequencies that he cannot hear.

Parti-coloured bat *Vespertilio murinus*, Sweden.

Common sheath-tailed bat *Taphozous georgianus*, Australia.

In the Australian tropics the arrival of the monsoon brings the first storm clouds for several months. The first rains mean the end of the heat and drought and the start of a new life. Plants sprout, termites and ants emerge and swarm and raise towards the sky in millions, to the enjoyment of the sheath-tailed bats that feed up there. Their young are born during the richest part of the year.

As soon as the pup is born, it clings on to its mother's hair or is enclosed by her tail membrane and wings. The pup then uses its special front milk teeth to affix itself to its mother's nipple.

Shortly after giving birth, the female must go back out on the hunt for food, usually leaving her pup back at the roost, if deemed sufficiently sheltered, where it sits on the wall or in some nook along with the rest of the colony's neonates, commonly under the watchful eye of one of the females. Sometimes, pups and females live separately since the young require a warmer roost, which gives the females a well-needed break. If, for some reason, a pup has to accompany its mother, such as when relocating to a new roost, it attaches itself by its teeth and feet to her belly keeping a firm grip on one of her nipples.

When a female returns from her hunt, she seeks out her pup by listening for its calls and then as it gets closer, by smelling her way to it.

Neonates have a personal cry that the mother easily recognises, just like a human baby. These cries are initially easily audible to humans, but as the pups grow, they become more high frequency. Each pup also has its own special scent that serves as a signal to its mother, again just like other mammals. The mother also lets the pup lick around her mouth, presumably in order to share her bacterial flora and minimise the danger of infection.

Pups are suckled both before and after a hunt, sometimes several times a night when very young. The females of certain tropical species are known to take their catch home to their pups, although this is probably extremely unusual. Vampires and nectar eaters can give their young blood and nectar, respectively, to supplement the milk they suckle.

The shorter the insect season, the more important it is for the pups to grow rapidly and become independent as soon as possible. Consequently, females choose warm roosts for as long as their pups are developing, as this promotes both growth and lactation. A temperature of around 35°C is ideal as it means that food and the females' energy reserves can be devoted to producing milk and speeding the growth of their young rather than to generating body heat.

Some females can produce almost their entire body weight in milk per day. It is during this period that her energy requirements peak, and her hunting skills are really put to the test. Often, she must hunt all night long. In a couple of species of flying foxes, the males also lactate, a peculiarity unique amongst mammals. However, it is not known if they actually suckle or if the milk they produce is a side-effect of the oestrogen content of their food.

Come high summer, it is time for the young to leave their nursery for the first time and learn to fly. The northern bat will serve to illustrate how this happens. At first, their trips are short and confined to the area around the nursery, but as the young grow stronger and become better fliers, they start to venture further away. Learning to echolocate and

Year after year the pregnant females return to the same place, the same church where they were born themselves long ago. The growing pile of droppings inside the window shows that the colony has resided in the same place during many years.

Pond bat *Myotis dasycneme*, Latvia.

catch insects will be a problem for a later date. To start with, the mother meets her pup at their exit hole, sometimes after having been out hunting for a while, and even if it might seem as if the pup is out flying unaided, the mother is never far away. However, a few weeks afterwards, the young begin to explore further afield and hunt insects for themselves. The mother remains nearby but not as close as before.

Before autumn turns to winter, the pups must have learnt to become proficient fliers and echolocators and built up sufficient fat reserves to survive until the spring. When learning where to find food and how to hunt, the young listen to the older bats and follow them as they go about their business. It is probably their mothers that show them where food is to be found.

However, the juvenile period is a dangerous one and barely half of the pups survive into their first year, as many crash-land and fall prey to predators during one of their first flying attempts. It is therefore extremely important for the females to choose nursery sites that are well out of the reach of cats, owls, and other predators and where their young can learn to fly in dark places beyond the glow of streetlamps and houses.

Mating

While the evenings are still warm and when the young have become sufficiently independent, the nurseries dissipate into smaller units and the females depart. The young normally stay for a week or more before spreading out over the landscape and migrating slowly towards their winter habitats. This is when the mating season for many bats begins. The way this happens differs distinctly between and even within species. A few species live in a monogamous relationship formed through a mutual courting procedure in which the two bats fly in circles. Other species are polygamous and engage in song-based mating rituals and territorial defence.

The African yellow-winged bat is one of the few species believed to live in monogamous relationships in which male and female help each other to raise and defend their pups. It is not known how stable or lasting the coupling is, but sometimes older pups remain within the group and form a mini colony together with their parents. The female and male always spend time together for a while in the afternoons, preening each other's coats in preparation for the evening hunt. As darkness approaches, they both hang from a branch waiting to pounce on passing prey.

Most bats are, however, polygynous or promiscuous, which means either that the males defend a harem or that no long-term bond is established between the sexes. A common mating strategy amongst male bats is to defend some kind of resource—say a good hunting ground, a roost, or a small group of females— from encroachment by other males. The strategy he chooses depends on how the females in the area behave, if they live in stable groups and how dispersed they are. Small, stable groups of females can be relatively easily controlled by a male and constitute a harem, which in some species can be sustained for a long time, in others not. Many bat species form harems, for example, tent-building leaf-nosed bats, of which the males defend both the home and the females. In some bat species that form large colonies, the colony can comprise several harems.

Male flying foxes normally control two or three females, and it is not uncommon for them to fight over them and the best sites. However, in areas where the females are dispersed and mobile, it is easier for the males to defend a feeding or roosting spot that the females use than it is to try to hold together a harem. In this way, they have direct access to the females over a small area that they patrol with special calls and behaviour patterns intended to deter inquisitive rivals.

For species that migrate over long distances, such as noctules, it is common for the male to claim tree hollows or other temporary

Two bat species live together in the old monastery. In the middle of the summer the young can already fly, but are recognised by the grey juvenile fur. Two species living in the same roost is not unusual, but they seldom form a single colony. These two species sometimes do so, however.

Greater horseshoe bat *Rhinolophus ferrumequinum* and Geoffroy's bat *Myotis emarginatus*, Portugal.

Brandt's bat is the commonest bat in the taiga. It lives in trees, cottages and barns, under tin roofs, asbestos and tiles. It is sometimes obvious that the colony has used the same roost during many years.

Brandt's bat *Myotis brandtii*, Sweden.

As darkness falls over the East African savanna the yellow-winged bats are already awake among the branches of the acacia tree. Their color provides camouflage for the bats during the day. They have good vision and excellent hearing, as suggested by large eyes and long ears. Yellow-winged bats live in pairs and defend a small territory, a somewhat unusual life style in bats. While hunting, they hang from a branch, waiting, ears move, listen. Sounds of wing beats are heard and rustling noise in the dark. Suddenly, a quick attack and one of the bats return to the branch with the prey. Like in other tropical bats the young are born at the beginning of the rains and the emergence of insects.

Yellow-winged bat *Lavia frons*, Kenya

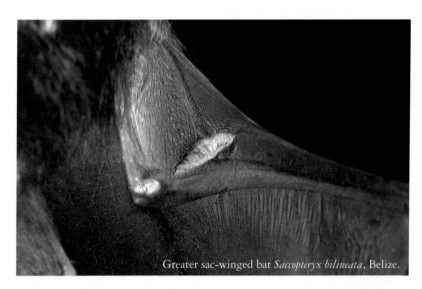

Greater sac-winged bat *Saccopteryx bilineata*, Belize.

The species in the family Emballonuridae are called sac-winged bats. They live in harems and the males mark their territories with scent and song-flight displays. The English name derives from the pockets on the male's wings, where the perfumes are stored.

residences along the route where he likes to hang, singing mating songs to attract females. When a female responds, he escorts her into the hole only to return shortly afterwards on the hunt for more. Sometimes he can be joined by another male attempting to gain access to the females. Mating then takes place in intervals during the day. Another migrating species in Europe, Nathusius' pipistrelle, exhibits the same behaviour, although seemingly with a more flexible approach. The males of this species are known to gather in groups close to the females' nurseries, tolerating each other's presence but chasing off inquisitive males from other groups. The mating calls of the Nathusius' pipistrelle peak at the end of July to August just before the females relocate for the season to their winter roosts. The closely related soprano pipistrelle begins at roughly the same time of the year, sometimes even earlier in June. The males patrol set routes close to their chosen roosts, announcing their presence with their

song in a territory that can stretch over a hectare or more. However, given that it is unlikely that the females, which are either pregnant or with neonate pups, would be interested in mating in June, such singing probably has other purposes.

The song of the parti-coloured bat is perhaps the best known in Scandinavia. The concert season starts after the first frosty nights in late September or early October, but on the other hand can continue on mild evenings well into winter. The singers patrol back and forth along rocky walls or tall buildings, the artificial cliffs of the city. Suitable roosts for the singer and any females he has can be found close to where he sings in crevices or ventilation pipes. The rock face or building also reflects the song and renders it audible over longer distances. To human ears it sounds like a series of four regular ticks a second, but in reality, it is much more complex. It is a familiar sonic feature of the urban soundscape in the late autumn, at least for anyone with acute hearing.

In West Africa, another spectacular song can be heard. Here, hammer-headed bats hang in rows, singing and flapping their wings in the hope of attracting females in an exhibition that is somewhat reminiscent of the mating behaviour of certain birds. The bat derives its name from the male's enormous head, which serves as a resonance chamber reinforcing its hooting mating call. Other African flying foxes are also good vocalists and perform a regular, sonorous song while impressing the females by exposing the white patches on their shoulders.

Many species lack any humanly discernible mating strategies. Probably the best thing for most is simply to mate when the occasion arises rather than trying to claim a territory and a harem. One such occasion is when the bats arrive at their winter roost, and it seems

Remains of a long dead industrial epoch hide deep inside a Central American rain forest. The sugar mill has been taken over by Nature and one of the valves is occupied by a male bat, every day he sits on exactly the same spot. It is the mating season and sometimes he is visited by females, but most of the time he is the only bat in the factory.

Greater sac-winged bat *Saccopteryx bilineata*, Belize.

Parti-coloured bat *Vespertilio murinus*, Latvia.

that the late autumn is the primary time for mating in northern latitudes. However, some species mate later during the winter and have taken to waking now and then from hibernation to do so. The males circle, select a female, and try either to wake her or, failing that, mate anyway while she is still asleep. A female can be visited by several males, which means that she saves the sperm from different individuals during the winter. This leads to rivalry between the males, but as usual it is the female that takes the ultimate decision. The female, or her eggs, seems able to distinguish chemically between different sperms, possibly allowing her to choose which sperm, and therefore which male, is to fertilise the egg. Female horseshoe bats seem able to make a more active choice and can form a kind of mucus plug to prevent further mating.

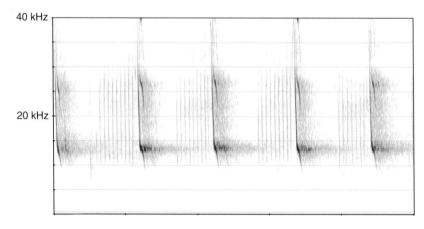

A male bat announces the mating time at Saint Peter's church in Riga. The parti-coloured bats have a special social life and the males are talented singers. The bats migrate from the country side to the cities, where the males let their songs echo from brick- and concrete cliffs, all to impress the opposite sex. The song is a familiar sound during mild evenings well into the winter, even in our city centres. As can be seen in the sonogram, the song is very different from any echolocation calls, being more complex and containing lower frequencies.

Swarming

As the nights grow longer, bats start to congregate outside cool mine entrances and cellars—first mostly males but soon also some females. Different colonies, species, and families agree on the same space, and all are there to take part in what is referred to as swarming. To a casual observer, it looks, perhaps bafflingly, as if the bats are flying around in circles chasing each other in and out of the hole. The bats are not there to hibernate, as they head off again before the temperature really drops. Instead, they hang around temporarily before flying on to other swarming sites. This swarming behaviour is probably a mating opportunity for individuals from a broad geographic area, but it is also a chance for older bats to show the young the location of suitable hibernation sites. However, the fact that relatively few females take part in this ritual contradicts somewhat these two hypotheses. Maybe it is nothing more complicated than that the bats use these cooler roosts to kick-start the hibernation process so that they can optimise their energy storage during the autumn nights. The question is then why the females do not have the same need. The purpose of swarming is still a mystery despite it being a spectacular and much-studied phenomenon.

Delayed Fertilisation

The gestation period for a bat is normally 2 months, although there is of course a wide deviation depending on the species and their habitats. A bat that mates in September could thus give birth late in November if there was nothing to prevent it from doing so. However, so that a pup will not be born at the wrong time of year, foetal development must somehow be postponed. This is most often achieved through delayed fertilisation, whereby the sperm deposited during mating is stored in the womb until the spring, when the bat ovulates.

The process is slightly different in the bent-winged bat family, but the outcome is the same. The egg is fertilised and implanted in the womb on mating, but does not start to develop until some months

While the late summer nights are still warm, bats gather at the old millstone mine. Mostly males in the beginning, later also females, different species at the same time, chasing each other in and out of the mine. They are all there to participate in the swarming. But the show is over already in September. Sweden.

Brown long-eared bat *Plecotus auritus*.

Daubenton's bat *Myotis daubentonii*.

later. There are other variations on the same theme, all designed to synchronise birth with access to food. The reason mating does not take place in the spring is that neither sex emerges from their long hibernation in sufficiently good physical shape nor there is simply not enough time for the pups to be born in time for the summer.

Because it is far easy for a bat to raise its pup to independency during a brief summer season, the females normally give birth to a single young once a year. However, in some species, such as the common noctule and soprano pipistrelle, twins are common. Some North American bats can give birth to as many as four young at the same time, although this is a rare exception. It is not only the short season that sets a limit on the size of the litter. Given the size and weight of bat foetuses, a large litter would compromise a female's manoeuvrability so much that she would find it hard to catch food during her long pregnancy and provide for her young once they are born.

Survival and Lifespan

To compensate for their small litters, bats live for a long time, usually 5–10 years but not uncommonly up to 20. Some bats are known to have even made it to over 40. Generally speaking, there is a relationship between maximum age and body size in mammals, whereby small animals with fast metabolisms rarely live more than a few years. Bats, however, live on average three, sometimes ten times longer than would be expected of animals on their size.

Roughly half of all young bats survive their first year, which is fairly normal for mammals. The chances of surviving an additional year become greater as the bat grows older and more experienced and skilled so that by the second year, they are often as high as 90%. In this sense, bats are more similar to large social animals, like primates, elephants, and whales, than they are to small ones like rodents and shrews. Bats live with their fellows for a long time, learning, sharing experiences, and giving birth to small litters that they spend a great deal of time and energy raising.

One reason bats live for so long is that they are very careful not to expose themselves to danger. Flying by night and hiding by day is without doubt the most important aspect of the bats' defensive strategy against the relatively few enemies they have. They are also highly cautious on the wing and make sure that where and how they hunt does not leave them vulnerable to attack.

Their lifespan may also be related to their metabolism. It was once thought that bats that hibernate for half the year live longer than their tropical cousins, which remain active for longer, owing to the prolonged drop in their metabolic rate. This is not the case, however, as tropical species appear to have similar life expectancies to those living in temperate climes.

One vital clue to their ageing process can be found in the mitochondria, the cellular organelles that convert oxygen and sugar to energy. As a natural product of the metabolic process, the mitochondria also release substances that can damage proteins and DNA. Compared with many other mammals, bats have few such substances in their bodies and can therefore, it is thought, enjoy a degree of longevity. They are also better able to resist disease, a topic we will be returning to later. The truth of the matter is that we do not really know why bats live for so long, but there is growing scientific interest in this intriguing problem, and we wait for the answers with bated breath.

Soprano pipistrelle *Pipistrellus pygmaeus*, Sweden.

Hibernation and Migration

BATS HIBERNATE AND MIGRATE as a way of dealing with declining food sources, regardless of which part of the world they inhabit. Most use both strategies in one way or another.

Winter Hibernation

The autumn is an important time of the year not only for mating but also for storing fat for the winter. Bats can eat a great deal in a short space of time, with some individuals almost doubling their body weight during August and September. When the autumn arrives, insects become inactive, leaving bats to survive on their stored energy reserves until the spring. In order to manage without food for several months, many small and even some large animals enter a state of hibernation, in which their metabolism enters standby mode and generates very little heat, reducing their body temperature to that of their surroundings. Mammals normally cannot survive a drop in body temperature of even a few degrees, but bats and some other species such as dormice and hedgehogs are exceptions. Small bats consume 1–2 g of fat during the winter, which corresponds to roughly a third of their summer weight.

The state of hibernation that bats enter is not to be confused with the winter dormancy of animals like bears and badgers, which is more like a prolonged sleep in which the body temperature drops to only a few degrees below normal and with no drastic reduction in physiological function.

Bat hibernation requires a reservoir of energy made up of fat tissue that is primarily stored over the back and shoulders. Much of this reserve is what is known as brown adipose tissue, funnily enough abbreviated as BAT. Unlike "normal" white fat, brown fat does not contribute to a cell's energy supply but only generates heat for use when the bat wakes out of its torpor. Brown fat is also found in newborn human babies and starts to produce vital body heat as soon as they see the light of day.

When a bat hibernates, its pulse drops from roughly 400 to between 20 and 40 beats a minute depending on the ambient and therefore the bat's own temperature. The brain and heart are the only organs to receive a normal blood flow during hibernation. Oxygen uptake drops by 99%, which means that sometimes bats breathe as little as once an hour. Minimal activity over a long period requires the regression of multiple neuronal synapses (connections) in the brain, which reduces the activity of the nervous system and its energy consumption. In theory, a bat can hibernate for 4 successive months, but in practice they occasionally wake up to pass water or faeces or to mate. Each awakening costs a tenth of a gram of fat, so those with modest fat reserves should not wake up too often. However, bats wake up more often during warmer conditions, which naturally lead to higher energy consumption. Some bats roosting inside a mine move to slightly cooler spots towards the end of the winter when their fat reserves start to run out. By far the greatest amount of stored fat is put into the actual awakening and subsequent flying. Hibernation itself consumes very little fat.

The soprano or pygmy pipistrelle thrives in urban habitat and has no problems to find suitable roosts in modern houses. This brick wall is home for a big group in winter. The picture was taken 22 November 2015, on the first real frost night. This was apparently the day when the bats decided to move to an insulated and warm winter roost.

© Springer International Publishing AG, part of Springer Nature 2017
J. Eklöf, J. Rydell, *Bats*, https://doi.org/10.1007/978-3-319-66538-2_7

If it is warm enough outside for insects to appear, bats can take the opportunity to stretch their wings and do a spot of hunting before returning to their winter roost. Some species wake more often than others, especially the soprano pipistrelles, which can be seen hunting winter crane flies and other small insects active on mild winter evenings as soon as the mercury rises above zero or, in extreme cases, even during the day. But the soprano pipistrelle is an exception. Most bat species in the north do not normally eat during the winter except when facing starvation and can manage without sustenance provided that they have found a secluded and sufficiently cool—but nonfreezing—winter roost.

A bat's preferred hibernation temperature depends on the species and on the size of its fat reserves, but usually it lies somewhere between 2 and 8°C. Some species, such as the barbastelle and the parti-coloured bat, prefer it colder but not so cold that they would need to burn fat for warmth. If it gets too warm, a bat's metabolism increases, generating more heat energy and causing a premature and potentially fatal depletion of its fat reserves. Choosing the right temperature for the winter roost is therefore extremely important for bats.

Earth cellars, the insulated walls of buildings, wells, stone bridges, mines, and scree make good winter roosts for bats in the north. Where the winters are milder, there is poorer access to cool places in caves and buildings, and bats might prefer to roost in hollow trees, which often provide the kind of cold environment needed to slow down the metabolism and minimise energy consumption. However, wood is a poor insulator, so the temperature in a hollow tree can fluctuate quite widely. To mitigate this effect, bats create their own insulation by clustering, a behaviour that they rarely exhibit when temperatures are more stable, at least in Europe.

But as well as being evenly cool, a winter roost must also be safe from human interference and predators, such as cats and martens and even field mice and shrews. Instinctively knowing they will be unable to fly to safety, bats make sure to find dark, secluded places in which to hibernate. If they are attacked, their only defence is to bare their teeth and hiss like a snake in the hope of frightening off whatever is disturbing their repose. The ideal strategy, however, is to hang somewhere out of the reach of such predators, like high up in the ceiling, in narrow crevices or hollows, or inside insulated walls.

When it is time to wake up, the bat's heart starts to beat faster, pumping blood through its layer of brown fat and slowly warming the body. Rebooting the brain requires a specific protein that helps to re-establish the retrogressed neuronal synapses. This protein is also found in the human brain but interestingly not in people with Alzheimer's diseases, which makes bat hibernation extremely interesting to this line of medical research. It takes about half an hour for a bat to wake from a deep torpor, heat its body to 40°C and get ready to fly. The process is extremely energy demanding, and as already noted, frequent awakenings during the winter seriously deplete fat reserves. When the spring arrives, bats emerge and head for the nursery, possibly with a few necessary temporary roosts on the way and the odd brief period of torpor when the weather is bad and there is no food to be had. Pregnant females also enter torpor during longer spells of poor weather to ward of starvation and to prevent the premature birth of the pup in the event of a late spring—before, that is, there is sufficient food to be found.

Hibernation in Warmer Climes

While months-long hibernation is mostly a characteristic of insect-eating bats living in temperate climates, it has recently been found that the inhabitants of the desert regions of the Middle East and even the tropics exhibit similar behaviour. More or less lengthy periods of deep torpor coincide with long dry spells or poor weather when insects are in short supply. In Hawaii, bats relocate from the tropical conditions by the sea to an elevation of almost 4000 m on the enormous volcano Mauna Loa, where they spend the winter in deep lava caves of a suitable temperature. In Taiwan, some roundleaf bats do the same. Come winter

The condensation of water in the fur on this hibernating bat shows that the bat is colder than the surrounding air. This is because it leans against the rock, which is even colder at the moment and therefore conducts heat away from the body.

Daubenton's bat *Myotis daubentonii,*

In Denmark good hibernation sites for bats, such as the limestone
mines on Jutland, may have tens of thousands of bats in winter, mostly
Daubenton's bats and pond bats. There is a massive spring emergence of
these species from the mines in April, when this picture was taken.

Pond bat *Myotis dasycneme*, Denmark.

Potato cellars were invaluable parts of every northern farm in former times and this is still the case in the Baltic countries. The bats hibernate over the roots, strictly protected by the owner of the cellar. Brown long-ears and potatoes have the same modest requirements in winter, dark and 2–5 degrees.

Brown long-eared bats *Plecotus auritus*, Estonia.

The bent-winged bats in tropical Africa hibernate, just like bats at higher latitudes. They migrate to limestone mines at higher elevation where the temperature is around 20 degrees, considerably lower than in the lowlands. This seems to be a viable strategy during the dry season, when insects are sparse and it may be necessary to save energy.

Bent-winged bats *Miniopterus* sp., Kenya.

The old water works at Tainan in Taiwan has been restored partly for the roundleaf bats that hibernate there. The energy saving is considerable even at about 20 degrees.

Formosan roundleaf bat *Hipposideros terasensis*, Taiwan.

they fly up to the mountains 2000 m or so above sea level, where they rest in the Second World War bunkers and tunnels that they have long since reclaimed. When summer returns, they fly back down the slopes to the lowland forests, where they give birth to and raise their young. The bats that stay in the lowlands all year round can also save energy by hibernating. Even though ambient temperatures reach around 25 °C, the bats can save 90% of their energy costs by hibernating, by use of the 15 °C drop in body temperature.

The bats of hot desert regions also benefit from hibernation to make it through periods when insects are scarce. Even though the sun may be blisteringly hot, desert nights are often chilly, allowing rocks and scree to cool down enough to remain at an acceptable temperature during the day. In deep caves and rocky crevices, it may actually be quite cool.

Migration

The ability to hibernate and live off stored fat means that bats can effectively stay in the same area all year round or at least avoid having to migrate long distances. Some species, such as the horseshoe bats and the brown long-eared bat, are also highly sedentary, rarely flying more than 20 km between nurseries, hunting grounds and winter roosts. Sometimes they even remain in the same roost all year round. They can live in spacious attics that provide necessary heat in the spring and summer and maintain a steady temperature of just above freezing in the winter. Bats in the north migrate neither to escape the cold nor to find a year-round supply of food, like birds do, but to relocate to suitable winter roosts. Consequently, they are equally content to migrate northward, and it is not unusual for them to fly vertically to find a layer of air at just the right temperature.

Some species choose to migrate long distances, sometimes as far as 2000 km in a single journey, despite the proximity of suitable places in which to spend the winter. However, this category of bat species is small and in Europe comprises only four or five species, two of which are proper long-distance migrators—namely, the common noctule and Nathusius' pipistrelle, both of which fly southwest in the autumn from Scandinavia and the Baltic countries to milder climes in Central Europe. Bats normally migrate in short legs of 1–2 h, breaking for a while to hunt and rest, although when flying over open sea they sometimes must fly for up to 8 h non-stop. Migrating bats have little trouble flying over the Baltic Sea, and the traces left by bats that have landed on oil rigs far out into the North Sea indicate that they also fly more or less regularly between Norway, Denmark, the Netherlands, and Britain. Sometimes bats can be seen arriving simultaneously at temporary roosts, suggesting they migrate in groups, at least from time to time.

What the long-distance migrators mainly have in common is their stamina in the air and their wings, biologically engineered for energy-efficient flight. Nathusius' pipistrelle, the common noctule, and the lesser noctule live in hollow trees, which in southern Europe are cooler during the winter than caves and buildings, while the parti-coloured bat inhabits crevices in buildings or rock faces. Temperature-wise there is little difference between spending the winter in a tree in Germany or a mine in Sweden. Bats are flexible in this respect, and the common noctule for one prefers to spend the winter in buildings in areas with cold winters, such as east Europe.

The likely advantage of migrating to warmer climes is that the earlier arrival of spring lengthens the breeding season and gives bat pups a better chance of surviving their first year. Ovulation occurs earlier in bats that have migrated south than in those that remain in the north, and presumably their young are born earlier as well. The birthing dates for the remainders vary by up to a month from year to year depending on weather and wind. It is known that pups that are born earlier in the

Some tropical bats die if the body temperature drops more than a few degrees below 37 and are unable to hibernate. This restricts their distribution to warm parts of the world, where food always can be found. But in Mexico the energetic budget in winter is near the limit and the bats migrate to take advantage of warm caves to save energy.

Leaf-chinned bats – Ghost-faced bat *Mormoops megalophylla*, Wagner's leaf-chinned bat *Pteronotus personatus* and Davy's naked-backed bat *Pteronotus davyii*, Mexico.

On 15 May 2012, a parti-coloured bat reaches the outer islets of the Swedish south coast. It is midday and the parti-coloured has travelled 200 kilometers non-stop from Poland. It is unusual that bats show up in the middle of the day, they usually move protected by darkness at night. But when the jump across the sea is too long, the migrants have no choice but to fly in sunlight towards the end of the trip.

Parti-coloured bat *Vespertilio murinus*, Sweden.

Nathusius' pipistrelle is the real long-distance migrant among the European bats. They fly thousands of kilometers between summer and winter homes and cross both the Baltic Sea and the North Sea. On the migration routes they stay in various crevices including bird- and bat boxes, alone or in small groups. Gunars Petersons studies the migration in Latvia, where they pass each spring and autumn on their way between Russia and Central Europe.

Nathusius' pipistrelle *Pipistrellus nathusii*, Latvia.

season fare better than those born later, as they are larger and more able to retain their weight during their first winter hibernation, reach sexual maturity earlier, and are generally healthier. They also reproduce more successfully later in life.

Long-distance migration is more a reproduction strategy than a survival strategy for bats, so males do not necessarily have to migrate to the same extent as females, which are the ones that need to time the birth of their young, optimise food resources, and follow the warm winds to the nurseries. All the males have to do is make sure they stay close to the females during the mating season. Otherwise they just need to look after themselves and can enter torpor whenever they need to. In some species, the sexes live more or less separately during the summer, so while the females fly north to feed, the males remain in their winter quarters all year round or migrate a little on the way. This means that in Sweden, for example, there can be few, if any common noctule males, while in southern Europe few, if any females. It is only when the autumn approaches that they gather at the swarming and mating sites along their migratory routes.

In the tropics, too, bats tend to follow the seasons and migrate between their summer and winter roosts. Mexican nectar-eating bats go to wherever the cacti and agave are in bloom and fly from the country's southern and central parts to the Sonora Desert in the north to satisfy their nectar requirements. Here too the long-distance migrants tend to be the females, and once they arrive in the north, the pups are born.

Navigating

Just like birds, bats make use of a magnetic sense to navigate over long distances. In one study, researchers released bats into an artificial magnetic field, and while initially able to alter the bats' flight path, it did not take long for the bats to realise their mistake and correct their course. The discovery of magnetic sensitivity in bats is a relatively new one, and we still do not know quite how they use this sense, which they presumably calibrate with the rising and setting of the sun. It was also recently found that bats can detect the direction of electromagnetic waves and make use of polarised light, the patterns of which, although dependent on the sun's position in the sky, are also visible after sunset. Just how bats perceive this light remains, however, a mystery.

Their sense of vision allows bats to pick up vital topographical clues and to distinguish subtle shifts in light, such as silhouettes on the horizon. Back in the 1960s, Donald Griffin, one of the two discoverers of echolocation, established that bats use their eyes to navigate and find their way home, using visual landmarks. It has also been demonstrated that bats, in theory at least, are able to see and navigate by the stars. However, given their poor focusing powers, it is possible that they fly by constellations and other patterns rather than by individual stars. Small bats use echolocation, of course, but owing to its short range, this is not a particularly effective means of navigating over long distances. Bats prefer to fly along riverbeds, treelines, and valleys, geographical features that are easily recognisable by more than one sense. Common noctules have been known to follow motorways, so maybe they are also able to navigate by using the streetlights.

Nevertheless, it is likely that many bats fly at high altitudes, just like birds, where their powers of echolocation can hardly be of any use. How bats navigate over longer distances is still largely unknown, and we still have much to learn about how their magnetic and visual senses interact.

Bat is called Fu, which also means happiness and prosperity with hope of a long and healthy life. In China it has long been understood that bats become unusually old and seldom are sick. It has also been realised that they live together with others over long periods, apparently in happy relationships. Bats are generally welcome in Chinese houses and occur as symbols everywhere, particularly in temples, as part of an active religion. They also occur on ornaments on bowls and jewels and as reliefs and artwork. They are not always easy to recognised but they are there for those that look carefully. Bats sometimes occur together with another symbol for Fu, a circle with straight lines and angles inside. In Taiwan, which the Chinese revolution never reached, Taoism is still an important part of life for millions of people.

Tsao-Ten Gon temple, Bein-Gan town, Taiwan.

Diadem roundleaf bat *Hipposideros diadema*, Thailand.

Roosts

For many people, the word "bat" conjures up images of dark caves, castles, and crevices. But bats find their home in all conceivable places, from subways to roofs, from termite mounds in the African savannah to palm leaves in the South American rainforest. Bats choose their roosts with great care and spend much of their lives in them. A roost provides not just a roof over the head but also a potential nursery, mating ground, and place of rest and hibernation. Above all, the roost provides protection against predators and the weather. Most bats also have several roosts to choose amongst should the one they currently occupy be disturbed. So, it can be hard for even the most experience bat watcher to find bats when they are not out flying. Many bats change their roost with the season, some relocate daily, and others return to the same place for decades. Roughly half of all bats use trees and other kinds of vegetation, the rest preferring buildings and caves or cave-like environments, such as rocky crevices, tunnels, mines, and cellars.

Natural Caves

In almost all bat families, there are species that live in caves, a preference that probably emerged early on in their evolution. Caves are permanent and have predictable and stable microclimates, with different spaces that offer the variations in temperature and humidity needed during the different seasons of the year. They are also dark, which

Caves are formed in limestone when the rock is dissolved by running water over long time spans. They are almost invaluable for bats and in karstic areas, such as in southern Thailand, the bat fauna is always very rich.

provides the perfect shelter from predators. There are not many predators able to catch a bat hanging from the wall or ceiling of a cave, although large centipedes and certain snakes can do it.

Genuine caves are formed in principle only where water runs through limestone for a long time and so are only found where the geological conditions are right. Even though there might be many caves in some areas, they are generally quite far apart and often far from suitable hunting grounds. On the other hand, the caves can accommodate many tenants at one time.

In Denmark, Daubenton's bats and pond bats have been found spending the winter in their thousands huddled up in scree inside the old underground limestone quarries. Northern bats seem to do the same under Norway's steep rocky cliffs. At these latitudes, any caves there are make excellent sites for winter hibernation, but they are far too cold to serve as nurseries in the summer.

When their young are small, they require a lot of heat. Eckert James River Cave in Texas is the destination for millions of bats, which fly there every summer to feed. Each day, several tonnes of droppings and urine accumulate on the floor, and the heat energy generated as all this breaks down combines with the body heat of the bats to raise the temperature in the cave and sometimes even make it dangerously stuffy. The urine can also cause very high levels of ammonia in the air, sometimes so much so that it bleaches the ends of the bats' fur. The stench of ammonia can be smelt for miles around, and anyone visiting the cave must wear a face mask. The bats, however, have mucosa that filter out the ammonia to prevent corrosive damage to their lungs. Not far from Eckert James River is Bracken Cave, which contains the largest assemblage of

© Springer International Publishing AG, part of Springer Nature 2017
J. Eklöf, J. Rydell, *Bats*, https://doi.org/10.1007/978-3-319-66538-2_8

mammals known. The number of bats there has been estimated at 20 million, and when they leave their roost every evening, the cloud they form reaches 40 km long and is clearly visible on radar. Every night, the colony eats at least 100 tonnes of insects, many of which are moths, including species considered pests to the farmers growing maize and cotton in the area. The bats therefore save them hundreds of thousands of dollars a year in pesticide costs alone. For this reason, the city of San Antonio has recently ruled that Bracken Cave and its environs be protected from all development and other human interference.

In Africa, too, cave-dwelling bats are of economic value to agriculture, but unlike in Texas, no one has calculated how great. However, there perhaps is an even more important effect, namely, that millions of bats reduce the number of mosquitoes and thus presumably help to keep malaria, one of humanity's most serious scourges at bay. In Malindi on the Kenyan coast, tens of thousands of bats congregate in caves, which with different microclimates in different chambers provide shared homes for many different species. Some prefer the dark, damp conditions deep in the caves, where they find warmth in narrow holes or high up in the domes of the ceiling. Others are at home in the relative cool air near the cave mouth and can sense how the light changes with the hours. It might seem extremely crowded and chaotic in the cave, but there is in fact order, at least in bat terms. Every individual returns to its own territory in the morning, sometimes to exactly the same place, year in year out. Some bats share a territory and bunch up on top of each other without it causing conflict, suggesting they are probably siblings or mothers and their young. The females must be able to pinpoint their young immediately, as there is not always time for them to search amongst the tens of thousands of hungry, screeching pups.

In southern Europe, bats can live all year round in caves, since the climate is sufficiently warm in the summer and sufficiently cool in the winter. The typical denizens of southern European caves are horseshoe bats, bent-winged bats, and greater mouse-eared bats. The horseshoe and greater mouse-eared are also found in central Europe, but there the caves are a little on the cold side in the summer, so the bats prefer to set up home in attics and eaves, which may be warmer. The Mediterranean is home to a rather odd cave dweller, the Egyptian fruit-bat, which is the only flying fox species in Europe. Normally, flying foxes live in trees and hang from branches, but this particular species has become adapted to a life in caves.

Ever since humans started to dig up the ground and build in stone, we have helped bats by creating new artificial caves. Bats have lived in our stone walls, pyramids, bridges, tunnels, mines, cellars, wells, barns, houses, churches, and castles. More than any other animal, they have followed in our footsteps, adapting themselves to a life in the cultural environments we create.

Mines and Trenches

In the Middle East, bats have seen civilisations come and go for millennia. In the ancient city of Petra in Jordan, horseshoe bats rest amongst the shadows of the carved-out cliff face, and in the pyramids of Egypt, mouse-tailed bats have been residents for at least 3000 years or so. In our modern times, we have also created new living spaces for bats, not least through our martial nature. Abandoned military bunkers along the River Jordan are an exclusive sanctuary for bats with a view of the water and the insect-rich hunting grounds there. The bunkers provide welcome shade from the heat and cover from predators. But they are also home to snakes that lie in wait for a careless bat to drop down from the ceiling.

During the Second World War, Hitler built a defence system to protect Berlin, the capital of the Third Reich, from invaders from the east. Ostwall, near the village of Nietoperek in Poland, consists of

The cave ceiling is roost for thousands of bats. Several species share the space in the cave, all having slightly different preferences. Some like the humidity in the cave interior or search for the heat trapped in narrow chambers and in high domes of the ceiling. Other species prefer cooler air near the cave entrance.

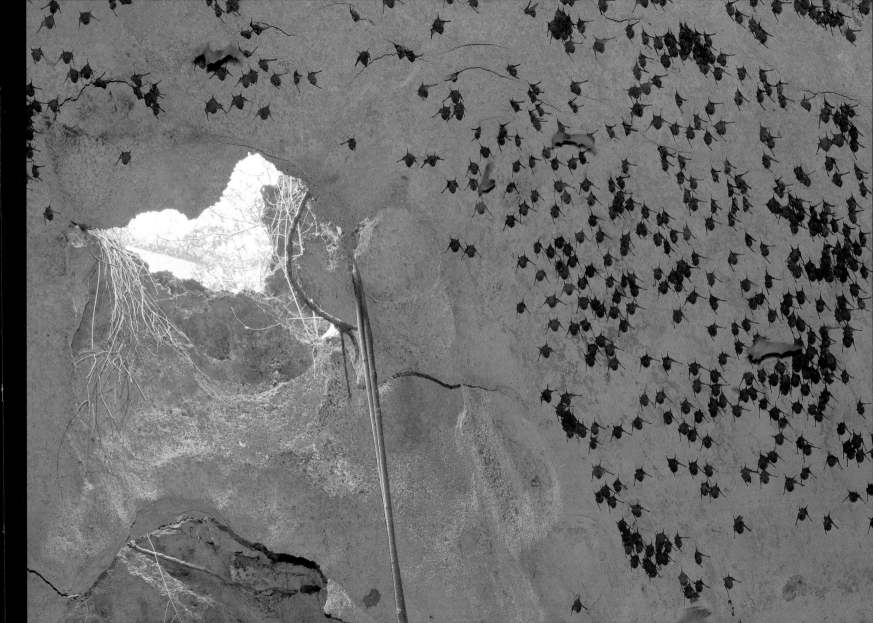

Egyptian fruit-bat *Rousettus aegyptiacus*, Kenya.

Cockroaches.

The cave ceiling is nursery, kindergarten, school and a place for social activities. The floor is grave yard and recycling station.

30 km of corridors connecting bunkers and other rooms buried 40 m underground. The facility was never used, however, and soon fell into oblivion. Instead, bats moved in and today 38,000 individuals from 12 different species hibernate in the tunnels. In the 1980s, the Polish president sought to improve the nation's economy by turning all the passageways into storage space for Europe's nuclear waste. However, as it turned out, the entire network of tunnels and the fields and wetlands above were awarded Natura 2000 status, and it is no exaggeration to say that Ostwall is now one of the continent's most important objects of nature conservation.

Nazi Germany's war machine was dependent on Swedish iron, and to supply the demand an old fifteenth-century mine, mentioned in the writings of Carl Linnaeus for its unique mix of fauna and flora from the Swedish mountains and Central Europe, was reopened. The area is now a nature reserve offering popular mine tours and a permanent bat exhibition. Nine species live in the vicinity during the summer and at least six hibernate in the mine tunnels inside the mountain. Daubenton's, Natterer's, and Brandt's bats hang side by side with northern and brown long-eared bats, while outside parti-coloured bats sing their songs in autumn and winter.

Bats are equally happy to use other cave-like spaces, especially the ones they find in old castles and fortifications. In present-day Latvia, the bridge linking Warsaw and Saint Petersburg over the Daugava River was a historical stage of the battle for supremacy between Europe and the Russian Empire. Here, in an attempt to stop Napoleon's troops, the Russians built Dinaburg, a castle that remained in military use until well into the 1990s. Here in its murky nooks and crannies, thousands of bats have found a winter roost. Renovation work is now underway at Dinaburg with EU money for the benefit of both its human and chiropteran visitors.

Shortly, before Napoleon's march on Russia, Sweden had lost Finland, prompting the construction of a fortification to which the Swedish king and the country's gold reserve could be transferred in the event of an invasion from the east. Karlsborg has now lost its

Human history is also the bats'. In the Middle East the sometimes violent history has created new space for bats. In military bunkers from the October War in 1973, bats have found exclusive roosts with view over the Jordan River.

Lesser mouse-tailed bat *Rhinopoma hardwickii*, Israel.

Greater mouse-eared bat *Myotis myotis*, Poland.

Forty meters under the ground in eastern Poland, near the village of Nietoperek, there is a system of bunkers and thirty kilometers of tunnels, built by the Nazis to defend Berlin from attacks from the east during the Second World War. But the system was never used. Instead the bats moved in and during the 75 years the number of bats has reached 38,000 in winter, distributed among 12 species. The bunker system and surrounding farm areas and wetland are designated as Natura 2000-areas.

status as Sweden's second capital but has been an alternative residence for the rare forest-dwelling barbastelle. The sun-warmed walls attract insects at dusk and protect hunting bats. With their muted echolocation, the barbastelles prey on moths amongst the parapets, and when autumn arrives, they head for the passageways and vaults to hibernate. The black coats of the barbastelles stand out against the pale castle walls until the dew drapes its glistening veil over them.

Bridges and Tall Buildings

Bats do not need castles or fortresses, however. Old stone bridges make excellent year-round homes for bats, which find protection and

Early April and most bats have already left the bunker for the season. But the mouse-eared bats remain for another few weeks, some stay through the summer, form a maternity colony and raise young well concealed under the ground.

In autumn the barbastelles move towards the old fortification tunnels, where they will spend the winter. The dark fur in sharp contrast to the light grey limestone walls. The fort is an important winter roost for barbastelles and a Natura 2000-designatied area.

Western barbastelle *Barbastella barbastellus*, Sweden.

The Dinaburg fortress was built where the road between Warshaw and St. Petersburg crossed the river Daugava, or Düna, in what is now Latvia. Today the fortress is a magnificent and somewhat frightening memorial of both the Tsar-empire and the Soviet-state. But in its dark corners bats in thousands have found a winter home.

Daubenton's bat *Myotis daubentonii*, Latvia.

The old iron mine at Taberg is used for hibernation in winter by several hundred bats of seven species.

Brandt's bat *Myotis brandtii*, Sweden.

Old valve bridges as well as modern motorway bridges sometimes harbor bat colonies, but the bats are usually hard to find. Rock and concrete store a lot of heat and provide protected roosting sites among stones or in expansion joints. Esthetic aspects are less important and this also applies to the trucks that pass within a few centimeters. This bridge is on the motorway near Veracruz in Mexico.

Gray sac-winged bat *Balantiopteryx plicata*, Mexico.

Churches are popular summer roosts for bats. A large and complex building like a church provides many accessible spaces with different conditions and under the roof it surely gets warm enough for the young. In this church two bat maternity colonies, serotines and pond bats, reside and Nathusius' pipistrelles and researchers sometimes pay visits as well.

Pond bat *Myotis dasycneme*, Latvia.

These bats eat other bats and are left alone in the well. Nobody wants to share roost with a *Cardioderma*-colony. In the dry season clean water is a desirable luxury for bats and humans alike. A deep well provides this to both and it also offers protection for mothers and growing pups in a cool roost.

Heart-nosed bat *Cardioderma cor*, Kenya.

Soprano pipistrelle *Pipistrellus pygmaeus*, Sweden.

In insulated house walls the bats find warm and protected roosts. The females help each other to raise the young, sharing knowledge and experience with young and old alike. The soprano pipistrelle forms large maternity colonies, sometimes with hundreds of females and their pups, which sometimes are twins as well. Nothing of this can be seen from the outside, however. The evening discussions and cries from hungry pups have to be appreciated through the wall, or by watching the intensive traffic at the exit hole under the roof.

uniform temperatures between their stones and a close source of food, especially for those that hunt for insects across calm watercourses. Larger, modern bridges can also provide a home for bats, since concrete and asphalt, once warmed, retain the heat for a long time, and all the bats have to do is find suitable cracks and hollows much like their rock-face-dwelling relatives. Deep cracks have a natural temperature gradient, so that how warm or cool a bat's surroundings are depends on how near the surface it elects to sit. In Sweden, parti-coloured bats often hibernate in rocky crevices although it is not uncommon for them to settle in the ventilation ducts of tall buildings, which are effectively the same thing, further down which they settle as the winter progresses. Sometimes, parti-coloured bats can even be found having relocated all the way indoors.

The noise and traffic of the city seem to cause little concern to bats, not even to those that live under road bridges. The main thing is the protection and microclimate these roosts afford, so it is no surprise to find bats living in distinctly urban environments, such as road tunnels, drainage ducts, and storm drains, the last of which also offer the advantage of combining protection with a source of water. In drought-stricken countries, where clean water is a desperately sought-after luxury, the local well becomes the hub of the community, human, and bat alike.

Attics and Basements

Sweden's long-eared bats know where to roost to get their food served to them. For centuries, they have occupied barns, attics, and church

A nectar-eating bat leaves the hollow in the Guanacaste tree. Her pup clings to the belly. Perhaps the mother does not dare to leave the pup at home or it has to learn where to find food. Tropical trees often have rotten cores. In the hollows bats reside and their droppings constantly fertilise the tree. The trees grow faster and set more fruit, and hence keep up well in the competition with other trees.

Pallas's long-tongued bat *Glossophaga soricina* and Guanacaste tree *Enterolobium cyclocarpum*, Belize.

towers, spaces that are sometimes suitable for both nurseries and hibernation. The bats clear attics of vermin and catch beetles, earwigs, bluebottles, and crane flies from ceilings and walls; their long ears listening patiently for telltale sounds – no scrape or rustle is too weak. Other species also use church buildings, and a little further south in Europe, it is common to find horseshoe, serotine, and pond bats hanging from the ceilings.

Other species are less particular about the aesthetics and are just as happy to live in houses under asbestos tiles or behind wooden facades. Bats are not gnawing animals or nest builders and therefore cause no damage to our homes. Instead, they use existing structures and nooks in interior walls, along roof beams, under tiles, or around chimneys and windows. They often seek out spaces where different materials meet, such as cavities between wood and stone, and that are often high up in as dark, sheltered, and warm environment as possible. Houses often provide many good hiding places at different temperatures and much better protection than trees, for example. During the summer, the females seek out heat and move in behind south-facing walls and under roofs, usually returning to the same place and the same colony where they themselves were born. Each colony has several reserve roosts to which they can return when the temperature or the disruption caused by vermin, predators, or humans becomes unbearable.

In the autumn, bats leave their summer roosts for, ultimately, their winter ones. For some species, this can mean flying across the garden to the nearest earth cellar, which was an invaluable feature of the nineteenth-century homestead as a larder for the family's potato and fruit harvests during the bleak winters. Here, bats could find a dark, frost-free winter residence, and some still do, provided the door has not rotted away, huddling together between the stones in the company of cave spiders, an oft-ignored feature of the Swedish cultural landscape.

In winter, the heated parts of modern buildings are far too warm for bats, although the insulation against the outer walls can provide good conditions and even a suitable temperature gradient. Such spaces are probably used much more than we know. It is very difficult to discover hibernating bats there since they are silent and sedentary, and usually inaudible from within.

Without doubt, access to houses and other buildings has made life easier for bats in the north. Hunter-gatherers colonised the Scandinavian peninsular as the glaciers retreated, long before the arrival of the forests or other natural preconditions for bat colonisation. During part of the Stone Age and throughout the Bronze Age, humans settled on a large scale and build well-insulated homes and in all probability for bats too. Humans and bats have shared homes in Scandinavia through rain and shine ever since.

There is a myth, especially widespread in Britain, that bats have been forced into buildings as a result of over-deforestation. This is a misconception. Human homes are the preferred solution for many species since they offer safer, warmer roosts and the opportunity to move indoors when the weather cools or when the bats need to get warm. It is likely that the species that exploit human habitations, such as the soprano pipistrelle and the northern, brown long-eared and parti-coloured bats, actually have the edge on those that stick to trees and rocky crevices.

Amongst the Tree Trunks and Foliage

Modern human civilisation has only existed for a fraction of the time that bats have lived on this planet, so it is, of course, only recently that they have been able to move into our buildings. Tree hollows are therefore the natural choice of roost for many bats, especially in areas without caves.

Hollows are formed in trees through the action of the fungi and insects that attack the damage caused by lightning strikes and woodpeckers. In the tropics, many trees, especially the huge rainforest species, gradually rot away from the inside. Bats soon move into the holes and fertilise the trees with their droppings, causing them to grow more quickly and produce more fruit. Trees that rot from the inside thus fare better in competition with other trees, so it pays to provide homes for others. Some woolly bats of Southeast Asia live in the rainforest's insectivorous pitcher plants, the potlike leaves of which contain water and enzymes that trap and digest insects. For bats, however, they make concealed parlours complete with en-suite sources of water and food. Like the tropical trees, the pitcher plants also benefit from the bats' droppings.

In their hunt for a roost, bats first check the suitability of a prospective site in terms of the protection and microclimate it offers. Some species make do with the space behind a piece of loose bark, while others prefer large, cool hollows. Some want a small entrance while others look for holes high up. Broad-winged species that use clutter-adapted echolocation can find homes deep in the forest, while high-flying bats prefer easily located roosts along the treeline or in tall trunks at the forest edge.

Some sheath-tailed bats sit completely exposed on tree trunks in the rainforest, often in a neat row at the same distance from each other every day, year after year, relying on their camouflage to keep themselves invisible against the bark. Yet they remain vigilant to every danger, constantly ready to let go and fly off whatever the time of day or night. Other forest dwellers sleep more or less openly amongst the foliage either alone or in small groups in the lush treetops high above the ground. They are normally well camouflaged and very difficult to spot.

The North American *Lasiurus*-species are powerful flyers and real long-distance migrants. Some of them spend the winters in the cloud forests on the slopes of the volcanoes in Mexico, where they find suitable winter temperatures. They always sleep in the open hanging in trees, where the camouflage makes them very hard to spot.

Hoary bat *Lasiurus cinereus*, Mexico.

A colony of leaf-nosed bats share a tent in Costa Rica's rain forest, five adults and three pups. The fruit-eating bats give life to the tropical forests, contributing to succession and diversity. They can fly ten kilometer or more each night in the search for a fig tree with ripe figs. Seeds rain over the forest where the bats pass, germinate and eventually become new fig trees.

Thomas's Fruit-eating bat *Dermanura watsoni*, Costa Rica

Flying foxes also hang in the open amongst the branches and leaves. Protection from predators seems to matter in the choice of roosts, why they often settle on small islands or in towns and villages where they are relatively safe from other animals, including humans. Proximity to food and water is also important, of course, but flying foxes would rather fly long distances than expose themselves to unnecessary risk. Trees in which large colonies live are often more or less defoliated, which makes social interactions between individuals, such as grooming, defending females, and rearing young easier. Their wings, which they use to shade themselves from the sun and shelter from heavy rain, serve as a fan, a blanket, a parasol, and an umbrella as required.

Some bats use leaves to make tents. By biting into the veins of palm and banana leaves, for example, they cause the leaves to fold into a shelter just big enough for a male and his harem. This behaviour is found in a score of species around the world but mainly in the leaf-nosed bats of South and Central America. Such tents are normally built and jealously guarded by the constantly vigilant males, which abandon them at the slightest disturbance.

Banana and related plants have leaves that, when developing, first form a tube and then a cone before unfurling properly. These formations are sometimes home to bats, some species of which are specially adapted to this type of residence. They have suction cups on their hands and feet and, unlike other bats, sit with their heads pointing upward. Each leaf serves as a suitable home for a day or two before opening fully, whereupon the bats are obliged to move. The frequent relocations are undoubtedly beneficial to the bats in terms of evading predators and also benefit the hygiene.

African sheath-tailed bat *Coleura afra*, Kenya.

Living Together

THE SOCIAL LIFE of bats is a complex and multifaceted one of constant interaction between males and females, between females and their pups, and between rivals, relatives, and other bats. Bats can form lasting friendships, they groom each other, they alert each other to danger, they form a united front against inquisitive owls, they steal and share food, and they look after each other's young. The strategies of co-existence are almost as many as there are bat species. While some form stable, monogamous couples, others live in colonies of millions of individuals. Some live in small family groups or harems that stick together throughout the year (sometimes year after year), others change their groupings according to season.

Gathering in large colonies has, of course, both its advantages and disadvantages. On the positive side, it enables bats to cooperate—for example, in keeping warm or cool, depending on the season—and to develop sophisticated lasting forms of collaboration. It also provides safety in numbers from circling bird of prey when leaving the roost at dusk. Furthermore, while it is the mother bats' job to teach their pups, sometimes siblings and other relatives step in to help raise and protect them. Another benefit of group living is that it allows young, inexperienced members of the colony to learn where food sources and temporary resting places can be found and how to fly safely by accompanying

their more experienced elders—even into adulthood. On the negative side, however, living in colonies exposes individuals to the risk of viral and parasitic infection and intensifies the competition for food.

Social Life

There is, however, efficiency in numbers. It has been shown that suckling mothers and their young expend less energy in large colonies than in small, as the denser the group, the greater the insulation and the steadier the temperature. This probably is one of the main reasons why females congregate in nurseries in the spring. Another advantage is collaboration. Close relatives, often older siblings or half-siblings, help to raise a mother's young, and even unrelated individuals in the same colony are known to assist each other. An adult normally stays close to the pups when it is time for the colony to go out hunting, and in some cases, the pups are gathered together to facilitate their supervision—a crèche for bats, as it were. Even though a female can carry a relatively large pup on the breast, she usually leaves it behind—otherwise the pup attaches itself firmly to her nipples and clings onto her fur with its feet.

The males can seem less helpful when it comes to caring for the young. This is a typical feature of mammals and is simply due to the biology of suckling. Instead, the job of the males during the summer is to keep away from the females and their young so, as to minimise competition for food and the risk of infection. In some monogamous species, however, the males take active responsibility and, in smaller harems, defend both the living territory and hunting ground with demonstrations of flying prowess, aggressive calls, and scent marking.

The cave looks crowded and chaotic, but there is actually law and order. Each individual bat returns to its own little territory in the cave ceiling, sometimes occupying exactly the same spot year after year. Some individuals share the territory with others and sit tightly together. Perhaps they are sisters or mothers and young. It is important to use the same spot because the relationships with the neighbours are well established.

© Springer International Publishing AG, part of Springer Nature 2017
J. Eklöf, J. Rydell, *Bats*, https://doi.org/10.1007/978-3-319-66538-2_9

Since bats live for a relative long time and often use the same roost year after year, they can form close bonds with their nearest kin, and larger colonies can contain sub-groups that provide security and a forum for exchanging information about successful hunts. It has been observed that species that relocate often do so with a few select individuals, commonly relatives, held together by the females, which make sure to include their daughters and granddaughters.

Bats often rub noses, possibly to learn each other's scent. Mutual grooming also helps the bonding process. We mentioned above that vampires share their food with less successful individuals; consequently, they keep a particularly close eye on each other, making mental notes of the altruistic members of the colony; while the selfish are immediately ostracised, the generous can count on being fed in reciprocation when in need themselves.

Communication

The more we learn about echolocation and social language, the more complicated it becomes. While much has already been learned about the sounds used in communication between females and their young and mating and territorial calls, more and more different communicative sounds are being discovered all the time. In recent years, it has been found that bats cannot only tell the members of their closest circle by their voices but can also hear the difference between those of different colonies, suggesting the possibility of dialects. One study carried out in Florida showed that free-tailed bats live in many respects as if in a musical and sing almost constantly, each song depending on the context and composed according to message, be it a warning, information, directions, or an invitation. The fact is that they use something akin to syntax and syllables in what is effectively a chiropteran language. The males sing a tune every time another bat passes their roost, an aggressive one for the males and a welcoming one for females. This can mean a great deal of singing for free-tailed bats since the colonies often comprise millions of individuals in different groups and permutations of harem, sex, and age.

Moreover, the roost as well as the size and composition of the colony vary with the season, which could also explain why such behaviour has evolved, the nature of the social context necessitating a flexible and complex communication system.

When humans listen to music while holding a conversation, the two hemispheres of the brain have their own tasks to perform, the right processing the music and the left the speech. It seems that the same is true for bats. So, while echolocating its surroundings and being bombarded with a constant influx of echoes, a bat can also listen to what its fellows are saying. Just like ours, the bat's right cerebral hemisphere is sensitive to small changes in frequency such as melody recognition, while the left deals with the more rhythmic hunting calls.

The disk-winged bats of Latin America that live in furled banana leaves relocate when the leaves open and become useless as living spaces. To keep the group together, they use special sonic signals to show their location. The other year, researchers found that the individuals sitting inside a rolled-up leaf use more complicated sounds than those flying outside because the leaf serves as a kind of megaphone. With no leaf to amplify its calls, a bat on the outside has to make more effort to be heard.

Vampire bats—and probably a great many other species—also use simple songs to keep track of each other and to navigate, the ensuing exchanges sounding a little like the duets performed by songbirds.

Competition

One disadvantage of living in a large colony is that it increases competition for food, and sometimes a bat will have to fly far from its colony,

Towards the end of the summer the maternity roosts become crowded, parasites become more prevalent and the females' responsibilities reduce as the pups grow. As soon as the young can fly the colonies start to disperse with the wind. The mouse-tailed bats travel in small groups, to find cooler roosts free from lice and flees. On the way, they find rest in the Roman ruins at Lake Galilee.

Greater mouse-tailed bat *Rhinopoma microphyllum*, Israel.

Sundevall's roundleaf bat *Hipposideros caffer*, Kenya.

to find undisturbed hunting grounds. Bats prefer to be alone when hunting and to not have other bats screeching in their ears, so many species have separate summer residences and hunt alone in their own environments. While the females hunt over lakes and beaches, the males stick to urban or cultivated areas. Ecologically, they behave like two separate species, which reduces the competition between them.

To hunt in peace, some bats establish temporary territories. Walk, for example, along a lit road in Sweden in the late summer, and the hunting territory of the northern bat is easy to spot. Staked out between street lamps, it is patrolled by the bat as it flies back and forth snapping up the moths attracted by the light. If the bat happens to encroach upon another's territory, an angry cry from its neighbour will soon send it back. And if it finds another bat aiming for the same prey, it can emit a sound that either scares the rival away or jams its echolocation signals.

Conversely, there are also examples of bats advertising to others where food may be found. Some leaf-nosed bats that normally live on fruit sometimes catch swarming ants, and when they do, they have been observed to call out to their friends to tell them where the insects are.

Parasites

Bats can share the same nook for decades, with several generations living side by side. But they are usually not alone. The walls are teeming with life after years of cohabitation, and bat bugs, fleas, lice, and parasitic flies thrive and spread from coat to coat. Mites are also very happy to settle on the naked skin of the face and ears, where superficial blood vessels are within easy reach. Parasites and bats have evolved together for millions of years, and most of the parasites confine themselves exclusively to bats, sometimes to just a single species.

There is usually no reason to cling on each other unless it is needed for thermoregulation. By maintaining a personal distance, fleas and other parasites cannot move as easily from fur to fur.

While the parasites are mostly harmless, they are, of course, extremely irritating, causing bats to spend much of their time grooming to rid themselves of them.

The rapidly growing multitude of skin parasites is one of the main reasons that bats often relocate from one roost to another, almost daily for some of the species that live in hollow trees. Those that live in larger spaces, such as buildings or caves, relocate to other parts as much as they are able. However, traces of parasitic activity, such as holes in the wings and blood clots on the skin, are common towards the end of the summer.

Infections

In a world full of viruses and germs, there is the ever-present risk of infection, especially in colonies containing hundreds of thousands of individuals. Bats are known for carrying a number of unpleasant diseases, to which they are sometimes resistant or immune.

The one disease that most people associate with bats is rabies, a viral disease that attacks the central nervous system. Rabies is found over large parts of the globe and in all kinds of mammals, although mostly in carnivores such as dogs, foxes, skunks, and raccoons. While bats, like other mammals, can carry classic rabies, in Europe, it is usually a specific chiropteran variety of the disease—the European bat lyssavirus. It was once thought that close to 10% of the world's bats were carriers, but it is now clear that this figure was a gross exaggeration and that it is way below 1%.

Compared with the danger posed to humans by rabid dogs, bats are a fairly insignificant problem, with the exception of South and Central America, where infected vampires can be a threat to livestock. The incidence of human infection has also risen in recent years in Latin America as forests are turned over to pastureland for beef cattle, making life easy for vampires by introducing a virtually ubiquitous abundance of food, on which vampire populations are thriving and proliferating. The strain of rabies carried by vampires is, however, not bat lyssavirus but classic rabies, a more infectious and therefore dangerous form, probably introduced by human activity through the

Bat wings often have hitch-hiking mites specialised for this somewhat unusual life style. In the fur live fleas, lice and bat-flies, sometimes also a special kind of bed-bug. All are wingless and feed on the bat's blood. The parasites are a constant pain for the host and one of the real drawbacks with the colonial life.

Daubenton's bat *Myotis daubentonii* with mite *Spinturnix* sp. and bat fly *Penicillidia* sp., Sweden.

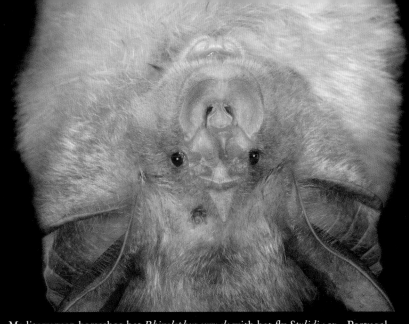

Natal bent-winged bat *Miniopterus natalensis* with bat fly *Nycteribia* sp., Kenya.

Mediterranean horseshoe bat *Rhinolophus euryale* with bat fly *Stylidia* sp., Portugal.

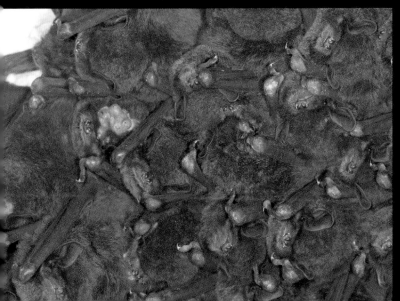

A hundred thousand mother bats return in the morning to the cave, where the new-born wait for milk. The females have filled their stomachs with nectar from the saguaro cacti, which flower in millions this time of the year.

These bats make long migratory flights across Mexico to track the flowering periods of different cacti and other plants. In early summer they arrive to this lava cave in the middle of the Sonoran desert, where they give birth and raise the young.

Lesser long-nosed bat *Leptonycteris yerbabuenae* and saguaro *Carniegia gigantea*, Mexico.

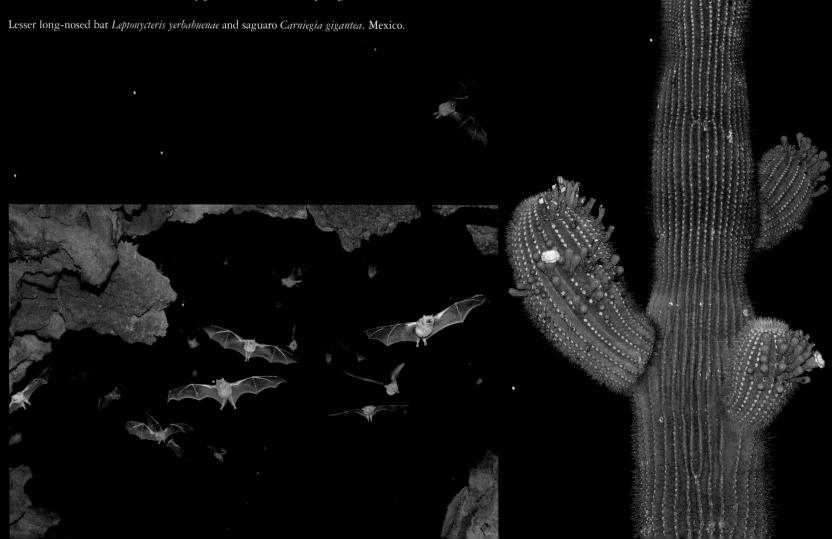

transportation of infected animals across the Atlantic. A number of misdirected interventions against vampires have merely aggravated the problem, and the poison and dynamite used to kill them off has only dispersed, rather than destroyed the colonies—with much collateral damage to many other species as well. In Mexico and some countries in South America, the problem has been overcome in many areas with the use of bovine rabies vaccines.

Bat rabies is, as noted, less infectious than classic rabies, and in some cases, the bats themselves are symptom-free. Yet the disease can still be transmitted to humans, so anyone handling bats—or any other wild mammal for that matter—should always put safety first and wear protective gloves. There are, however, effective vaccines that work even if taken several days after being bitten.

Even though there have been only a few incidents of humans being bitten by rabid bats, the fear and media fearmongering is widespread in Europe and North America. In 2014, however, articles about bat rabies were overshadowed by another virus, Ebola. The epidemic in West Africa is the hitherto most severe outbreak of this disease and is rumoured to have been started by a boy playing with an infected flying fox. Horseshoe bats living in a nearby tree were also blamed. The fact is, however, that no one knows if these bats were indeed infected. It has long been known that bats can act as a host to viruses able to cause different types of haemorrhagic fever such as Ebola, but it does not appear that the bats themselves are particularly susceptible to the disease, so it is hard to detect the presence of the virus.

Immunity and Resistance

Apart from Ebola and rabies, bats are more or less resistant to a wide range of parasitic and viral diseases, such as malaria, influenza, and the much written-about lung disease SARS. Some people argue that bats carry disproportionately many virus strains compared to other animals. In one species of flying fox, for instance, scientists have found 55 different virus types, 50 of which were previously unknown. While this is not unique for bats, and while similar finds can be made in many other mammalian groups, if one looks closely enough, there has been a strong focus on bats in recent times, which goes at least part way to explaining the many virus finds.

Whether or not bats carry an unusually rich viral flora, they are highly resistant to infection. There are also indications that bats rarely, if ever, develop cancer. Researchers recently examined mutations of 6000 genes and proteins during viral attack on bats and compared the results with human control cells. They found that the bat cells reacted much more quickly and effectively than the human, implying that their biochemical defence is simply superior to ours. The reason for this is a matter of speculation, but it could have something to do with their ability to fly. Active flight is extremely energy demanding and pushes the bat metabolism to levels way above resting. A number of by-products are thus formed in the body during flight, including DNA-damaging free radicals. Over the millions of years of bat evolution, their cell repair system has been fine-tuned to handle such damage as swiftly and effectively as possible. A fortunate side-effect of this could, then, be that other cellular damage is dealt with in a similarly efficacious manner. Perhaps, this is where we might also find the key to why bats live so long. It goes without saying that these properties of viral resistance, tumour prevention, and longevity are of great interest to humans and to medical research.

White-Nose Syndrome

One might be tempted to think that bats are immortal and that there is a possible grain of truth to all the myths, legends, and witches' brews claiming that bat blood can clean wounds and cure everything from blindness to a broken heart. There is, however, one disease that bats are powerless against, a fungal disease that affects the muzzle. It is called white-nose syndrome or WNS for short.

The disease was discovered in the late winter of 2006 in Albany, New York, by a speleologist who came upon a number of dead bats in a cave and took pictures of individuals that seemed to have a mysterious

Being alone can be advantageous in some cases. The risk for infections is kept to a minimum. Single males are much less affected by parasites than females in colonies.

white mass around their nose and mouth. The full extent of the discovery was only truly realised, however, when more such white-nosed bats were found in the area the following winter and a proper investigation was launched. Researchers then found hundreds of dead bats and observed peculiar behaviours. For example, the bats were not hibernating normally but waking up from their torpor and flying around both inside and outside their winter roosts, and not only at night. For a bat with a winter fat reserve, this is, of course, potentially fatal and many of the dead individuals had simply starved. Since then, the problem has quickly escalated. In 2008, the first cases were found outside New York State, in Vermont and Massachusetts, and there were reports the following year from four more states. The disease now exists from Canada to Mississippi and has claimed 7 million lives from a dozen different species to date. In Pennsylvania's vast hibernation caves, only droppings remain of the once enormous colonies that scientists estimate have now been decimated to just 10% of their original size.

WNS is caused by a fungus that thrives in cold, damp environments—the same conditions that bats seek for their winter roosts. The fungus works its way in under the skin and infects an area around the mouth, nose, and wings, causing fatal behavioural abnormalities during hibernation. Bats, which normally resist everything from malaria to Ebola, seem to have found their match in this resilient fungus. But there is hope. The first unaffected colonies have recently been found in Vermont in an otherwise heavily infected area.

As with rabies, the fungus was introduced to North America by humans, possibly via the dirty boots of some cave diver. Indeed, the fungus is found in caves and mines in northern Europe, but does not show the same symptoms, as the bats here seem to have evolved counterstrategies in a way that their North American cousins have failed to do. Perhaps, it is to reduce the risk of infection that European bats usually hibernate alone or in small clusters, rather than the enormous congregations found across the Atlantic.

African sheath-tailed bat *Coleura afra*, Kenya.

Ghosts and bats

In Kenya the ghosts are still alive. Some of them are found in caves and are connected with bats. Such things are taken seriously and are important aspects of rural life. Caves are used for religious purposes, for communication with the ghosts, but for this the bats are also needed and must therefore be protected. Typically, part of the cave is used for worship while the rest is left alone, so in some sense religion is the bats' best protection. In this crevice below the boulders, bats and villagers share the space. Even Jesus resides in a corner.

Kit Mikayi Rock near Kisumu, Kenya.

Ghost-faced bat *Mormoops megalophylla*, Mexico.

Under the Cover of Darkness

ONE OF THE most characteristic features of bats is the life of secrecy they live under the cover of darkness. There is nothing unusual about mammals being nocturnal. On the contrary, it is a normal state, and bats have likely been night animals ever since the Cretaceous. Nevertheless, one might wonder why not a single species has evolved to be diurnal. What is it that has so consistently prevented bats from being active during the day?

The answer is undoubtedly small, swift birds of prey, but even birds such as crows and gulls occasionally hunt bats. The notion that bats have trouble dealing with such predators is corroborated by the fact that species living on islands where there are no birds of prey, such as the Azores, are much more active during the day than their mainland kin. But Azorean bats are also mostly nocturnal, so the daylight hours are not entirely safe there either.

Migrating bats sometimes must fly in the late autumn, which puts them in something of a tricky situation as by then the nights are too cold for any insects to be out and about. So, to find sufficient food for the long journey they have to take, they hunt in the low afternoon sun amongst swarming crane flies and other insects. For the same reason, bats are sometimes seen hunting in daytime in the early spring, but this must be considered as an emergency option. It is dangerous to hunt with sparrow hawks and other birds of prey nearby.

Shadows and darkness are the bats' best friends. At dusk even a small hedge or some bushes make a difference by providing shade and cover for the bats during their flight between the roost and the feeding grounds. Protected by the bushes, the bats can leave the cave earlier and get access to more of the insect swarms at dusk.

The bat flying technique and senses have been refined for a nocturnal lifestyle. Bats are slower than swallows and swifts, for example, and they have poor forward-planning faculties as their echolocation is so limited in range. An echolocating bat purposefully directs its call straight ahead to obtain as clear an echo as possible from its prey. It also means, however, that the bat has little idea of what is going on either side of and behind it. Moreover, its visual acuity is of low resolution and probably quite unreliable for detecting danger approaching at a distance. All this is the price the bat has to pay for being able to hunt at night using echolocation while having sufficiently sensitive nocturnal vision to navigate in the dark. The payoff is access to insects, protection, and freedom from competition from birds.

But of course, there are birds that also hunt at night. Owls are nocturnal, just like bats, and occasionally pose a serious threat. It is not uncommon to find bats in the pellets regurgitated by barn and tawny owls. Once an owl realises that it can hunt bats as they fly to and from their roosts, it can snap up quite a feast there. But the bats do all they can to defend themselves and their colony from potential devastation. For instance, we have observed a group of northern bats collectively chasing away an overly inquisitive tawny owl.

Leaving the Roost

Mosquitoes, nonbiting midges, and many other true flies briefly swarm at dusk before settling down for the night. Bats that live on such prey, such as the soprano pipistrelle and common noctule, therefore, have to

J. Eklöf, J. Rydell, *Bats*, https://doi.org/10.1007/978-3-319-66538-2_10

Black-bearded tomb bat *Taphozous melanopogon*, Thailand.

An owl couple with young shared a cliff with a little colony of bats. In the moonlight the owls waited for the bats to emerge, a typical hunting technique for owls.

African spotted owl *Bubo africanus*, Kenya.

be out in time if they are to feed, but not so early that they might still be visible to birds of prey.

The members of a bat colony leave their roost at almost the same time after sunset every evening, but the exact time differs from species to species and from roost to roost depending on how shaded its

The colony gathers inside the cave entrance before emergence at dusk. Nobody wants to be the first out, the predators wait for those about to leave into the darkness of the night. The bravest or hungriest ones, those that leave before the others, often fall victims, as well as the last.

location. Bats that live on small swarming insects leave relatively early in the evening, often while it is still quite light outside. Other species wait until it is safe and have to make do with other, less abundant prey, such as moths and beetles that fly later at night. Water-trawling species are usually particularly cautious and refrain from leaving the cover of the trees for the open water, where they are exposed to danger from above and perhaps also from below, until it is completely dark. In Finland and northern Sweden, Daubenton's bats abandon open waters completely around midsummer, when the nights are at their lightest and most perilous.

African sheath-tailed bat *Coleura afra*, Kenya.

Small bats that live on fruit and nectar in the tropics are also in no hurry to leave their roosts in the evening and prefer to wait until it is completely dark, having no reason to take unnecessary risks as their food remains available to them all night long. The same applies to vampire bats—the cattle they feed on are also conveniently sedentary. But flying foxes, which rely on their powers of vision, have to leave their roost at dusk while there is still enough light to hunt.

Before bats leave the safety of their cave or culvert, they fly around for a long while to assess how light or dark it is outside. Many bats congregate at the exit, unwilling to be the first out and the victim of a waiting predator. But as soon as one decides to take the plunge, they all do, each protected by a surrounding host of living shields. From the cavernous homes of enormous colonies, bats can emerge like pillars of smoke and disappear into the distance, dispersing high into the air once the immediate threat has passed.

Under the Midnight Sun

There are few bats in the land of the midnight sun. It is often too cold at night for there to be enough insects active and the summers are far too short for bat pups to grow up before the night-time frosts kill off their prey. Another problem in the north is that the nights are so short and light that hunting becomes perilous. Northern bats, which nevertheless live and reproduce at the Arctic Circle, can evidently handle the situation, at least in certain years. They hunt for a couple of hours in the middle of the night when it is at its darkest, sticking close to the shadows or the branches of trees. Up in the north, reproduction is delayed

making sure that the young are born in August when the nights start to get darker. The mother bats' need for food increases as the hunting periods lengthen, but the most important thing is that the young are spared having to learn to fly in July when the nights are still light, which would fatally expose them to predators.

In the Moonlight

It is said that bats are lunophobic and avoid hunting in the moonlight. It is more likely, however, that it is the insects they feed on that are not particularly active when the moon is bright. Insect-eating bats are not bothered even by a full moon, but this is still something of a moot point, and it seems as if species behave differently in this respect. Indeed, it might seem somewhat odd that the moon has little effect on bat behaviour given how bright moonlight can seem, but this is probably because birds of prey like sparrow hawks have much less sensitive night vision than both bats and humans, so even the brightest of moonlit nights is too dark for them to pose a threat to bats.

In the tropics, there are birds of prey and owls that are specialist bat hunters and that presumably see better at night than the northern species. There are also often snakes and spiders lurking by flowers to pounce on fruit- and nectar-eating bats. The leaf-nosed bats of Latin America therefore seem to take extra care when the moon is bright. The large flying foxes of the Old World have hardly anything to fear from spiders, but need to be on the lookout for snakes and themselves exploit the moonlight so that they can remain active into the small hours.

Everybody at the same time is the best strategy. Confidence is to have others around, a protective shield of conspecifics that are taken first.

Exploration of caves, mines and other sites underground have become a popular occupation, but it is not always compatible with the bats' need to roost in peace. Disturbance of bats at maternity- and hibernation roosts is illegal throughout Europe, but important localities also need protection by strong gates that allow bats to get in and out but stop humans without proper keys.

Natterer's bat *Myotis nattereri*, Sweden.

Of Bats and Men

ABOUT HALF OF all bat species are on "red lists", which means they are considered threatened or almost threatened with extinction. While every country draws up its own red list, there are also global and regional lists, for example, in the EU. Bats are as vulnerable to the rapid depletion of the earth's biological diversity as other plants and animals, and as usual the worst affected areas are in the tropics. Threats include deforestation, the industrialisation of forestry and agriculture, pesticides, polluted water, wetland drainage, the spread of disease, foreign species invasion, and climate change. These threats are fairly general. They affect far more plants and animals than just bats and are therefore discussed in a wider context far beyond the scope of this book. Instead, we will address issues that affect bats more specifically, whether negatively or positively, but that are largely overlooked by the usual nature conservation discourse.

Chemicals

Central Europe's populations of horseshoe bats, barbastelles, and pond bats declined dramatically in just a few years in the late 1950s, leaving caves and mines once populated by hundreds or thousands of bats desolate. This also happened in the southern USA. Since the 1930s, the modernisation of agriculture had brought about various changes to the environment, with more uniform landscapes and declining insect populations. It was, however, the introduction of industrial-scale pesticides in the 1950s that posed the greatest threat. Bats ingested PCB, DDT,

lindane, and other chemicals that were used to combat insects, protect wood, and clear brushwood, and being ravenous eaters, they can consume high doses of chemicals if they are present in their food. To make matters worse, the pesticides and fungicides in treated wood meant that their homes were also sometimes contaminated. However, these pesticides have since been banned, and the decline in bat populations has been reversed. In 2014, the European Environment Agency (EEA) compiled data for 16 different bat species from 6000 overwintering sites around Europe, and since the 1990s the number of bats has risen by over 40%. Some species are still threatened and rare, but most of Europe's bat populations now seem to be stable or thriving. The same trend was seen amongst birds of prey, such as the peregrine and the white-tailed eagle, which, like bats, also took about half a century to recover from the catastrophe. At the same time, it should be mentioned that these conclusions only apply to species that overwinter in large numbers in mines and caves, and that can easily be counted. The statistics do not include species that live alone or in small groups in trees or houses.

Disruption

There is another, less frequently mentioned reason for the decline in bat populations. In the 1930s, researchers began to mark bats with metal rings, as with birds, in mines and caves to study their movements and lifespan. Such projects multiplied in number and scale and were mostly conducted in the winter when the bats were hibernating and could be

© Springer International Publishing AG, part of Springer Nature 2017
J. Eklöf, J. Rydell, *Bats*, https://doi.org/10.1007/978-3-319-66538-2_11

picked from the walls by hand and tagged before being released. In the 1960s, researchers discovered to their astonishment that the bats had abandoned many of their regular winter roosts but were blind to the connection between their disappearance and their being woken from hibernation in winter. Roughly a million bats were tagged like this in Europe, mostly in Germany. These days it is deemed wholly unacceptable to disturb hibernating bats, and it is prohibited throughout the EU. For a few decades of the twentieth century, ring-markers could be the bats' worst enemies along with agricultural and forestry chemicals. Today, bat tagging is only done on a small scale in specially licensed research projects that keep the animals under close observation. This kind of tagging causes no distress to the bats and teaches science a great deal about their movement patterns.

However, the presence of intruders in caves and mines is a serious and rapidly spreading concern. Despite all the metal gates and heavy-duty locks, people break in and disturb roosting colonies, causing severe distress to countless individuals. There are caves in which most of the global population of a particular species are gathered and where regular disturbances are potentially devastating.

Illumination

Towns, industrial zones, and motorways are expanding at an ever-increasing pace, depriving built-up areas of proper darkness and turning night into dusky day. Bats, which for millions of years have sought refuge in the dark, are naturally affected by an ever more illuminated world. Nowadays, light pollution as a serious environmental problem has become a subject of growing attention in Europe.

Some bats completely avoid the glow of the lights, and for them a row of street lamps, for example, forms a barrier that fragments their hunting grounds. Horseshoe bats in Britain provide a much-studied example of this phenomenon, but mouse-eared and long-eared bats react in very much the same way. These bats very probably avoid artificial lights to minimise the threat from owls and other birds of prey but perhaps also because it interferes with their night vision. A bat's vision is at its optimal in the low light of dusk.

Artificial illumination is increasing outside towns and cities too. It has become something of a fashion to train spotlights on facades, bridges, and, above all, old churches to light them up like some Las Vegas in the Nordic night. To us it might look pretty and comforting, but to bats it is life-threatening. Brown long-eared bats that once populated every church steeple must seek out darker refuges out of sight from birds of prey, whiskered bats that used to hunt along the facades have lost their hunting grounds and Daubenton's bats have to take detours beside the illuminated bridge archways.

Other bats are drawn to the lamps. Light attracts insects, so the lamps advertise the availability of food for the swifter bat species, which have less to fear from owls. Examples of this are the northern bats of Scandinavia, Kuhl's pipistrelle in the Mediterranean area, and many African species of house bats. It seems very likely that these bats have the lamps to thank for their recent growth in numbers and spread, which might even have occurred at other species' expense. In Switzerland, the lights erected along Alpine valley roads have been observed to invite soprano pipistrelles, which invade and eventually chase out the horseshoe bats. Horseshoes never hunt around lamps, so there is probably some other form of rivalry between the species at work here. Lighting does not scare away all bats but means that a few species invade to the detriment of others. The bat fauna is changing, and the diversity becomes severely reduced.

When it comes to attracting insects, not all light is the same. It is primarily the ultraviolet (UV) component of artificial light that does the job and explains why light is also irresistible to certain bats. UV light is emitted by all the old mercury vapour lamps and by some of the orange-hued street lamps currently used along urban roads. The new LED technology is more flexible, and it is possible to avoid the wavelengths that are particularly hazardous and unnecessary, like

The urbanisation in Europe and elsewhere brings an ever increasing "need" to turn the night into day. Dark areas may eventually become islands in an increasingly lit landscape, like Vesuvius surrounded by the suburbs of Napoli.

A flood-lit wall attracts insects, which attract bats, particularly the open-air species such as noctules and pipistrelles. However, the lights repel other, more light-averse species of bats and may have a strong negative effect on the local insect populations. The flood-lighting of buildings is becoming a serious environmental issue. Sweden.

The old stone bridge has been renovated and equipped with aesthetic LED-lights. Unfortunately this has scared off Daubenton's bats, which no longer use the bridge for roosting. They do not even fly through the arches anymore, as they most likely have ever since the bridge was constructed in 1780. Sweden.

UV. Consequently, the problem of insect attraction can be reduced drastically, at least in theory. On the other hand, LED bulbs use little energy and are inexpensive to use, so there is a clear risk we will have even more of them, leading to escalating light pollution. For the light-aversive species, it doesn't matter what kind of light is used, even if certain bats seem to handle orange light slightly better. For these bats, all visible light is equally bad in principle, and they are likely to face increasing problems as urbanisation progresses.

The lamps also affect insect behaviour and reproduction, which could herald a biological disaster. Moths hatch at the wrong time in lit up areas, and the females produce less of the scent they use to attract males. Fewer insects not only mean less food for bats but also have many other consequences.

Apart from certain swift-flying insectivores, flying foxes can also benefit from more light. In Israel, date farmers curse the burgeoning population of bats there in the ever more artificially illuminated landscape. These visually guided bats normally seek out food at dawn and dusk and under a brightly lit moon, thanks to the use of lamps, they have little difficulty finding gardens and orchards and can feast all night long if they so wish.

Wind Power

The shift to fossil-free sources of electricity has led to a rapid expansion of wind power over the past decade and a half or so. The turbines have proved, however, to present a serious, often fatal threat to a great many bats. Whereas it was once thought that bats more or less accidentally collided with the turbine blades during migration, it is becoming increasingly clear that they actually seek them out in the hunt for insects and are killed either when struck by the blades or by the lung damage

Davy's naked-backed bat *Pteronotus davyii*, Mexico.

This little leaf-chinned bat hunts for insects around street-lights in Mexico. Its technique is very similar to that of the pipistrelles in Europe, which also feed around street-lights in this way.

Darkness is an important resource for bats. The church in Jeleniewo in Poland is still dark at night, despite its location in the middle of the town, and three colonies of bats share the space in the church – serotines, Nathusius' pipistrelles and pond bats.

Serotine *Eptesicus serotinus*, Poland.

In this church everything is as it should. No lights and the long-ears provide the pest-control in the tower and attic – for free.

Brown long-eared bat *Plecotus auritus*, Sweden.

Modern wind turbines have turned out to be dangerous to bats, which are killed when they collide with the moving rotors. Some of the open-air bats, such as noctules and pipistrelles, often fly at such heights when feeding or migrating and are therefore more often killed than other bats. For some migratory populations of bats the increased mortality at wind turbines is so high that their long-term survival may be at risk. This picture shows one of the old wind farms in Tarifa in southern Spain.

Serotine *Eptesicus serotinus*.

caused by the sudden changes in air pressure as they rotate. Fortunately, scientists can now pin down the problem and identify at-risk species as well as sensitive areas and times. The high-flying and migrating species are especially vulnerable, and at least in northern Europe, 90% of accidents occur during mild, calm late summer nights. This is when the bats start to leave their summer roosts to fly along their habitual routes to their mating and hibernation areas. Insects gather at a high altitude moving along stable layers of air pursued by hungry bats. So, it can be predicted with a fair degree of accuracy which turbines are dangerous to bats and which are not. An effective and fairly cheap way of minimising the danger is to keep the rotors switched off on calm, warm nights so that bats can hunt in peace. By carefully planning where to erect turbines and keeping them inactive during certain weather conditions, power companies can do much to mitigate the problem.

Bat Boxes and Cellars

A roosting box can bring you into close contact with bats. If you put one up near a source of food (e.g. by a lake) but far from a natural place to roost, you will soon find it occupied. The best thing is to set up several boxes in close proximity but facing different directions so that the bats have alternative accommodation when the need arises. Hang them in an open location, such as on a tree trunk, with an opening (ideally) at the bottom narrow enough for the bats to fly in and out but not birds. Otherwise, feel free to experiment with size, shape, and material.

It has recently been noted that bat boxes in Europe mostly attract soprano pipistrelles and sometimes common noctules, which have been observed to knock out other species that inhabited the area before they were erected. It is possible that roosting boxes are like street lamps that attract some common species, such as the soprano pipistrelle, at the expense of the more exclusive species, thus compromising species diversity. The value of the boxes might actually run contrary to their intended purpose and result in more soprano pipistrelles but fewer barbastelles and other rarer bats. This is not to say that you should refrain from setting up roosting boxes, as it might well be worth attracting bats to your garden, but you should be aware of the effect they can have on other species. Different boxes are suited to different species, and a careful choice of box might reduce the risk of it only benefiting the common species. It should be pointed out, however, that bat boxes can never replace the natural holes and cavities found in old forests and buildings.

Abandoned earth cellars can be renovated without risk and without inviting soprano pipistrelles, which never roost in such places. On the other hand, they are important winter roosts for the Scandinavian brown long-eared bat, but only if they are in fairly good condition with a sound door and a useable exit/entrance hole. Such cellars were found in the grounds of every farmstead and were a vital part of the nineteenth and twentieth century home before the arrival of fridges and freezers. Restoring old, dilapidated potato cellars is therefore a laudable contribution to culture and nature conservation.

Legislation

All bats in the EU are strictly protected by a law that prohibits the killing or harming of bats or their roosts and at certain times of the year they may not even be disturbed. If the law is obeyed, it can have many important consequences. For example, the facades of churches and other buildings may not be lit up without permission, and the same may apply to many other places. Wind turbines, roads, and homes may not be built without consideration to bats nor may buildings be demolished or renovated until it has been established that they are not inhabited. And perhaps the greatest challenge of all, due care must be paid to bats by the agricultural and forestry industries.

Ecosystem Services

It is not just for aesthetic and nostalgic reasons that we should care for bats and protect them from lit motorways and poorly placed wind turbines. There is an economic aspect as well. A few years ago it was

Mangroves colonise tropical sea shores and build new land in the tidal zone. Particles sediment around their stilt roosts and form a substrate extremely rich in nutrients. Fish, mussels and cray fish occur in abundance, supporting a rich and sustainable fishery. Intact mangroves also provide efficient protection against tsunami-waves. In Thailand much of the original mangrove has been destroyed but in Satun in the far south there is still a large area left intact. The mangroves flower at night and are pollinated by a little nectar-bat that seems specialised for this job. Hence, there are distinct connections between bats, a rich fishery and protection against tsunami-waves.

Lesser long-tongued nectar-bat *Macroglossus minimus* and white mangrove *Sonneratia alba*, Thailand.

Many bats adapt rapidly and easily to new conditions and this flexibility helps them to survive and prosper even in heavily exploited areas. But there are also many examples of the opposite. Specialised tropical forest bats stand no chance when their habitat is cut down and converted to plantations. Kakamega forest in western Kenya is still a protected oasis in an otherwise densely populated and heavily exploited area.

Bate's slit-faced bat *Nycteris arge*, Kenya.

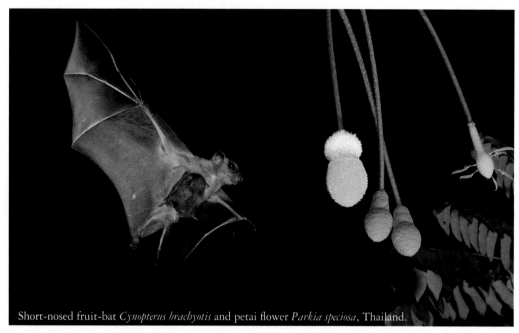

Short-nosed fruit-bat *Cynopterus brachyotis* and petai flower *Parkia speciosa*, Thailand.

Bats provide ecosystem services worth hundreds of millions of dollars annually in Thailand, as they probably do in every tropical country. Some pollinate fruit trees and other plants while others control pest insects – all for free. The stink beans are tasty and somewhat smelly and a peculiarity in South-East Asian cooking. The flowers are pollinated by bats.

calculated that bats contribute ecosystem services worth 20 billion dollars in the USA in pesticide savings alone—and this does not include the cost of the harm done by such chemicals to humans, animals, and the local environment. Insect-eating bats eat not only agricultural pests but also mosquitoes, presumably helping to keep diseases such as dengue fever, West Nile fever, and malaria in check. What this ecosystem service is worth has yet to be calculated. Since bats are not infected by normal malaria, they can provide important information about how to handle this human scourge more effectively.

Earlier in the book we saw how bats form a vital part of the tropical ecosystem by spreading seeds and pollinating flowers. The spectacular African baobab tree is almost invaluable to the savannah, and it and the animals that live in it all depend on bat pollination. Bat pollination of the durian trees of Thailand and Malaysia is worth over 100 million US dollars every year, and the durian is just one of many bat-pollinated fruit trees in Southeast Asia. Bat droppings, which are rich in phosphate and nitrogen, are used as fertiliser and are almost 50 times more expensive than common cattle dung.

Calculating the value of ecosystem services is a relatively new field, and we have really only just begun to understand the true value of bats, but in recent years there has been growing scientific and popular interest in them. Their immune system is studied to teach us more about how to prevent or cure new epidemics. Their wing construction, which once fired the imagination of Leonardo da Vinci, is used by engineers to create flying robots. Knowledge of echolocation and ultrasound is used to construct aids for the visually impaired and to develop self-navigating vehicles, and there are many high-tech innovations in ultrasonic diagnostics, sonar, and radar that bats have inspired. The once much maligned bat has been lifted out of the shadows, and we have only just begun to realise just how much it can teach us. It is time for the Western view of bats as creatures of evil and Gothic mystery to be superseded by the Asian, which regards them as holding the key to success and longevity.

In the earth cellar

In the Baltic countries the potato cellar is an essential part of every rural household. The cellar at Umbsaadu farm near Viljandi is maintained with great care, like many other cellars in Estonia. A small group of brown long-eared bats spend the winter in one of the ventilation shafts. The owner Mr Helijut Lepikult explains that this has been the case every winter since 1950, when he moved to the farm. The ventilation shafts are closed from the outside in November or December, when the arrival of the bats has announced the first harsh frost nights and the arrival of winter. The shafts are opened again in March or April, when the sun starts to heat the air and insects emerge. The bats leave their winter quarter, but will surely be back again next autumn. During the five or six winter months, the bats are locked in without access to food or water, but this is evidently not a problem. Likewise, they know that the daily visits to the cellar by the owners, when the door is opened and the light turned on for a while, is nothing to worry about

Brown long-eared bat *Plecotus auritus*, Estonia.

Hodgson's myotis (Formosan golden bat) *Myotis formosus flavus.*

Japanese pipistrelle *Pipistrellus abramus.*

Mr. Chang

Heng-Chia Chang had just started his new job as a teacher at the little country school in Yunlin. He used to walk across the school-yard watching the children playing football. Nobody then knew which creatures lived in the trees above. One day Mr. Chang was hit by a ball on the back of his head. He turned around to see where it came from. This was when he first saw them, the little golden bats hanging under a leaf in a tree above. Teacher and children gathered and soon found more bats in nearby trees. Mr. Chang became interested in the bats and initiated some research. He was told that the bats were very common in the area in the past, but that they have become much rarer now. He created an exhibit about the golden bats at the school and before long not only the children, but also artists and scientists, were engaged. No more than a couple of years after the bats were found, the school had evolved into an information center and meeting place for conservation-minded people in Taiwan, with plenty of visitors. The school has been decorated with bats in all forms. The children are proud ambassadors for bats and their conservation. In the bat boxes on the wall of the school house several breeding groups of another species, the Japanese

The Bat Guardian

Beatrice Amadalo proudly shows the diploma that certifies that she has been nominated as a "Bat Guardian" by the IUCN Species Survival Commission. Beatrice runs a farm together with her husband in western Kenya. The farm is a seasonal home to a large colony of flying foxes, sometimes as many as 40,000 individuals. To let the bats stay on the farm means a considerable cost for Beatrice and her husband, mostly because they risk conflicts with fruit-growing neighbours that would rather see the bats somewhere else.

Next to Beatrice is Paul Webala, while the flying foxes are counted by Beryl Makori. Paul is head of "Kenya Bat Research Project" at Karatina University and Beryl is his student. Well educated, conservation-minded people, like Beatrice, Paul and Beryl, are essential keys to the future in the tropics.

Straw-coloured fruit-bat *Eidolon helvum.*

Arabian horseshoe bat *Rhinolophus clivosus*, Israel.

Greater mouse-tailed bat *Rhinopoma microphyllum*, Israel.

Greater bent-winged bat *Miniopterus inflatus*, Kenya.

Cape long-eared bat *Nycteris thebaica*, Kenya.

FURTHER READING

Altringham, J. D. 2011. Bats: From evolution to conservation. Oxford University Press, Oxford.

Barataud, M. 2015. Acoustic ecology of European bats. Muséum national d'Histoire naturelle, Paris.

Dietz, C., von Helversen, O. & Nill, D. 2009. Bats of Britain, Europe & North-west Africa. A & C Black Publishers, London.

Dietz, C. & Kiefer, A. 2016. Bats of Britain and Europe. Bloomsbury, London.

Fenton, M. B. & Simmons N. B. 2014. Bats—A world of mystery and science. University of Chicago Press, Chicago.

Griffin, D. R. 1958. Listening in the dark. Yale University Press, New Haven.

Gunnell, G. F. & Simmons, N. B. 2012. Evolutionary history of bats—Fossils, molecules and morphology. Cambridge University Press, Cambridge.

Kunz T. H. & Fenton M. B. 2003. Bat ecology. University of Chicago Press, Chicago.

Middleton, N. Froud, A. & French, K. 2014. Social calls of the bats of Britain and Ireland. Pelagic Publishing, Exeter, U. K.

Russ, J. 2012. British bat calls—A guide to species identification. Pelagic Publishing, Exeter, U. K.

Rich, C. & Longcore, T. 2007. Ecological consequences of artificial night lighting. Island Press, New York.

Thomas, J. A., Moss, C. F. & Vater, M. 2004. Echolocation in bats and dolphins. University of Chicago Press, Chicago.

Voigt, C. C. & Kingston, T. (eds) 2016. Bats in the anthropocene. Springer-Verlag, Berlin.

www.africanbatconversation.org – African Bat Conservation
www.batcon.org – BCI, Bat Conservation International, USA
www.bats.org.uk – Bat Conservation Trust, U. K.
www.eurobats.org – EUROBATS

Kuhl's pipistrelle *Pipistrellus kuhlii*, Italy.

THANKS

Friends and colleagues in many countries have provided invitations, help and advice, logistic support, and permissions, as well as access to their study colonies in caves and houses. Without this backup, almost nothing of this project could have been achieved. We would like to express our sincere thanks to Damian Milne and Thomas Madsen (Australia), Luisa Rodrigues, Hugo Rebelo, Helena Santos, Pedro Alves, Bruno Silva and Silvia Pereira Barreiro (Portugal), Ernst Herman Solmsen (Costa Rica), Brock Fenton and his colleagues and students and the personnel at Lamanai Outpost Lodge (Belize), Eran Amichai, Arjan Boonman, Ivo Borissov, Ofri Eitan, Yossi Yovel, Carmi Korine and their students (Israel), Simon Musila, Paul Webala, Aziza Zuhura, Robert Syingi, Mike Bartonjo, Beryl Makori, Simon Masika (Kenya), Tomasz Kokurewicz and his team of students (Poland), Gunars Petersons, Jurgis Suba, Alda Stepanova, Viesturs Vintulis, Ilze Brila (Latvia), Raphaël Arlettaz (Switzerland), Chen-Han Chou, Heng-Chia Chang, Hsi-Chi Chen, Hong-Chang Chen, Kuang-Lung Huang, Hsue-Chen Chen (Taiwan), Antonio Guillén-Servant and his students, Anglelica Menchaca, Rodrigo Medellin and his students (Mexico), Javier Juste, Carlos Ibañez, Sonia Sánchez Navarro, Juan Quetglas, Domingo Trujillo, Rubén Barone (Spain), Bert Wiklund, Anita and Lee Hildsgaard Rom (Denmark), Jeroen van der Kooij, Keith Redford, Kristoffer Böhn (Norway), Matti Masing (Estonia), Danilo Russo and his students (Italy), Gareth Jones, Roger Ransome and the staff in Hereford Cathedral (England), Sara Bumrungsri, Tuanjit Sritongchuay, Kanuengnit Wayo, and C. E. Nuevo Diego (Thailand).

For logistical support during photography at home (Sweden), we like to thank Björn Arkenfall, Maria Bajuk, Ingemar Beiron, Sven Frändås, Magnus Gelang, Ingvar Hermansson, Thomas Persson, Lars-Eric Roxin, and Ola Rydell and for the help with things other than photography also Anders Blomdahl, Katharina Dittmar de la Cruz, Erika Dahlberg, Torbjörn Ebenhard (Uppsala University), Bengt Edqvist (Department of Dialectology, Onomastics and Folklore Research, Göteborg), Jenny Eklöf, Anders Hedenström (Lund University), Olof Helje, Henryk Hörner (Kleva mine), Espen Jensen, Catarina Krång (Wildlife Rehab Centre Stockholm), Sabine Lind and Hans Fransson (Taberg mine), Magnus Lindqvist (Tropikariet Helsingborg), Stefan Nyman, Björn Olsen, Stefan Pettersson, Charlotte Wedelsbäck and her family, Jan Westin (Universeum Göteborg), and Claes Wistberg

Last but not least, a big thanks to all owners of houses, castles, mills, barns, cellars, ruins, wells, gardens, and many other places where we have been allowed to work in Sweden as well as abroad.

Desert pipistrelle *Hypsugo ariel*, Israel.

SPECIES PHOTO INDEX

JOHAN EKLÖF received a Ph.D. in zoology at Göteborg University in 2003. He is now a zoologist and copywriter, working with bat surveys, conservation biology, research, and scientific communication. He has written fictional as well as non-fictional books on bats, animal evolution, and life in general. He travels throughout Sweden, not just in search of bats but also to learn more about folklore and bat myths. In 2016, he received a scholarship from the Swedish Writers' Union to explore the field of bat mythology.

JENS RYDELL received a Ph.D. in ecology at Lund University in 1990. He is a Swedish scientist and nature photographer, working with bat ecology and conservation at Lund University. Jens is the author of many scientific papers on bats and insects, recently with focus on the effects of wind power and artificial lighting on bats. He was awarded the Royal Swedish Academy of Sciences conservation prize 2017 and in the same year he reached the final in the Wildlife Photographer of the Year competition in the UK.

Johan and Jens have worked together with bats for more than 20 years. Both live near Göteborg, Sweden, but the bats take them all over the country and sometimes to other corners of the world as well.